TK 7868 .L6 B73 2001
Braga, Newton C.
CMOS sourcebook

CMOS Sourcebook

by Newton C. Braga

NEW ENGLAND INSTITUTE
OF TECHNOLOGY
LIBRARY

CD-ROM included in back of book.

Other PROMPT® books by Newton C. Braga

Sourcebook for Electronics Calculations, Formulas, and Tables

Fun Projects for the Experimenter

Electronics for the Electrician

CMOS
Sourcebook

by Newton C. Braga

©2001 by Sams Technical Publishing

PROMPT© Publications is an imprint of Sams Technical Publishing, 5436 W. 78th St., Indianapolis, IN 46268.

All rights reserved. No part of this book shall be reproduced, stored in a retrieval system, or transmitted by any means, electronic, mechanical, photocopying, recording, or otherwise, without written permission from the publisher. No patent liability is assumed with respect to the use of the information contained herein. While every precaution has been taken in the preparation of this book, the author, the publisher or seller assumes no responsibility for errors or omissions. Neither is any liability assumed for damages resulting from the use of information contained herein.

International Standard Book Number: 0-7906-1234-8

Library of Congress Catalog Card Number: 2001091655

Acquisitions Editor: Deborah Abshier
Editor: Cricket Franklin
Assistant Editor: Kim Heusel
Typesetting: Kim Heusel
Indexing: Kim Heusel
Proofreader: Kathy Murray
Cover Design: Christy Pierce
Graphics Conversion: Christy Pierce
Illustrations: Courtesy the author

Trademark Acknowledgments:
All product illustrations, product names and logos are trademarks of their respective manufacturers. All terms in this book that are known or suspected to be trademarks or services have been appropriately capitalized. PROMPT® Publications and Sams Technical Publishing cannot attest to the accuracy of this information. Use of an illustration, term or logo in this book should not be regarded as affecting the validity of any trademark or service mark.

PRINTED IN THE UNITED STATES OF AMERICA

9 8 7 6 5 4 3 2 1

Contents

SECTION 1

CMOS BASICS 1
Introduction 3
Digital Electronics Fundamentals 4
The Postulates of Boolean Algebra 31
Boolean Relationships 43
The CMOS Family 54

SECTION 2

FUNCTIONAL DIAGRAMS AND INFORMATION FOR DESIGNERS 87

DEVICES 90
4000 93
4001 95
4002 97
4006 99
4007 101
4008 103
4009 105
4010 107
4011 109
4012 111
4013 113
4014 115
4015 117
4016 119
4017 121
4018 124
4019 127
4020 129
4021 131
4022 134

4023	137
4024	139
4025	141
4027	143
4028	146
4029	149
4030	152
4031	154
4034	157
4035	160
4041	163
4042	165
4043	167
4044	170
4046	173
4047	176
4048	179
4049	182
4050	184
4051	186
4052	189
4053	192
4066	195
4069	197
4070	199
4071	201
4072	203
4073	205
4075	207
4076	209
4081	211
4082	213
4089	215
4093	218
4094	221
4099	223
40106	225
40160	227
40161	230
40162	233
40163	236
40174	239
40175	241

Contents

40192	244
40193	247
4503	250
4510	252
4516	255
4511	258
4512	261
4514/4515	264
4518	267
4520	270
4522/4526	273
4528	276
4529	279
4538	282
4541	285
4543	288
4723	291
4724	294

SECTION 3

Basic Blocks Using CMOS ICs 297

- Oscillators 299
- Monostables (One Shot) 323
- Bistables and Counters 336
- Complete Applications 370

Index 385

CMOS Sourcebook

by Newton C. Braga

SECTION 1
CMOS BASICS

Introduction

Although many parts of the modern circuit can be designed with complex circuitry containing hundreds, thousands, or even millions of components inside dedicated blocks, there are functions in which simple blocks are needed. In those cases, the advanced or high technology can be replaced by intermediate technology because why use an atomic bomb to kill a fly?

This is the case with the CMOS logic circuits. Although one can find equivalent functions by using a microprocessor, microcontroller, or DSP — there are large numbers of applications where this type of technology is not necessary. Easy to use, with very low power requirements, CMOS circuits are used to perform simple functions where simple functions are needed and to complete or interface with complex circuits or when needed. CMOS ICs are used often and will be used even more in the future.

A CMOS gate consumes only a fraction of microamperes of more complex devices and can be used with power supplies in a wide range of values. The large number of functions found in the CMOS 4000 family allows a designer to create nearly all digital circuits imaginable. But to know how to use CMOS ICs, it is necessary to have some basic information about digital electronics, the electrical characteristics of the devices of the 4000 family, and precautions to take when handling these components. It is also important to know how these ICs can be interfaced with other components of the same family and other families of components, such as transistors, operational amplifiers, SCRs, and Triacs.

The aim of the first part of this book is to give you basic information about logic, CMOS configurations, and ways to use these ICs. Because the book is intended as reference for designers, the information is limited to the essential. If you would like more information about theory and the components, look for specific literature or the data books of the components.

Digital Electronics Fundamentals

Some basic information about digital electronics is provided here. The information is brief because the aim of this book is not to teach digital electronics.

Digital Logic

The possibility that a two-valued number system can represent any quantity is the basis for the most powerful and sophisticated computers. Digital logic can explain how this works. For computers, everything is based on the binary number system. Only two symbols are used — 0 and 1 — to represent any quantity. We can specify, for example, that

- 0 = L = Low = no
- 1 = H = High = yes

Using zeroes (0's) and ones (1's), every statement or condition must be either true or false. It is not possible to be partly true or false. This approach may seem limited to some readers, but it actually works and can be used to express very complex relationships among any number of individual conditions.

One reason for basing logical operations on the binary number system is that it is easy to design circuits that can switch between two defined states with no ambiguity. Another important reason is that it is possible to build circuits that will indefinitely remain in one state. This is a very important fact allowing the construction of devices that can remember sequences of events.

Conventions

In this book the number "1" is used to represent the high logic level, or +V (the power supply voltage), and the number "0" is used to represent the low logic level, or ground. Digital logic can be divided into two classes:

CMOS Basics

- **Combinational logic** — The outputs are determined by the logic being performed and the logic states the inputs are in at that moment. The combinational logic is formed by functions, such as inverters, AND gates, NAND gates, OR gates, NOR gates, Exclusive OR gates, binary addition blocks, Multiplexers, and Decoders/Demultiplexers.
- **Sequential logic** — The output depends either on the logical inputs or the prior states of the outputs. The sequential logic is performed by functions such as RS NAND Latches, Clocked RS Latches, RS Flip-Flops, J-K Flip-Flops, and D Latches.

Other conventions:

- Vdd – to positive power supply
- Vss – to ground (0 V)
- Vee – to voltage
- A, B, C, D, etc – to inputs of logic blocks
- X, Y, W – to output of blocks
- Q – to output of flip-flops and bi-stable elements
- /Q – to complementary output of bi-stable elements
- CLK – clock
- CLR – clear

CMOS Logic

The CMOS logic is based on the use of complementary MOS transistors to perform logic functions with almost no current requirements. The typical CMOS gate needs only 0.01 mA for normal operation when powered from voltage sources between 3V and 15V. Starting from simple configurations, complex functions can be performed and used in CMOS ICs. The following are the basic configurations of combinational logic functions using MOS transistors and found in the CMOS ICs:

Inverter — The basic configuration of a CMOS inverter circuit is shown in Figure 1.1. The same figure shows the symbols adopted to represent this function.

CMOS Sourcebook

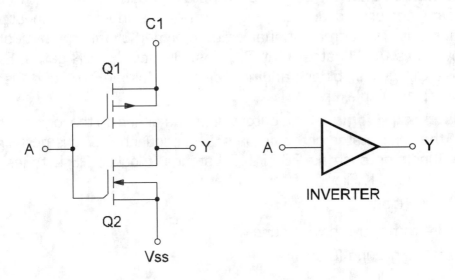

[Figure 1.1]

Both transistors are enhancement-mode MOSFETs: one N-channel and the other P-channel. The gates of these transistors are connected to form the input. The drains are connected together to form the output.

When the transistors are off, their resistance rises to many Giga ohms (almost infinite, for all practical purposes), and when on, their resistance falls to about 200 W. When they are open (on), no current is drawn. When in operation, the voltage in the output depends on the voltage applied to the input. If the input is connected to ground (0), Q1 conducts and the output voltage is +V (1). If the input voltage is +V (1), Q2 conducts and the output goes to ground (0).

NOR gate — Figure 1.2 shows the basic configuration of a NOR gate using MOS transistors.

This circuit shows that the transistor can be associated in series and parallel to perform the desired function.

CMOS Basics

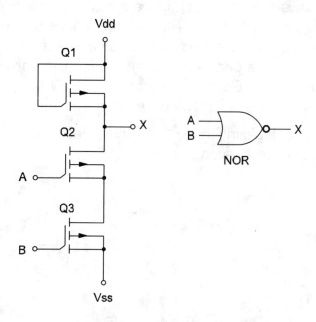

[Figure 1.2]

NAND Gate — The basic structure of a NAND gate using MOSFETs is shown in Figure 1.3.

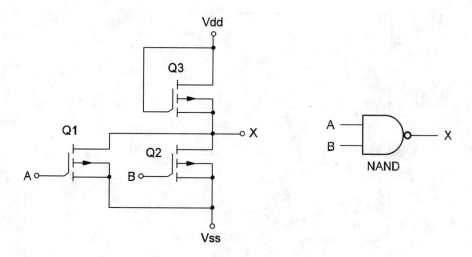

[Figure 1.3]

The circuit shows the basic configuration of a 2-input NAND gate. One limitation of this circuit is the combined resistance of the MOSFETs when in series. This problem is solved by the use of buffered or B-series CMOS gates. Other functions can be performed with the combination of these functions in the same ICs. For instance, an AND function can be made from the association of a NAND gate with an inverter.

For the sequential logic in Figure 1.4, some basic configurations found in CMOS applications are given.

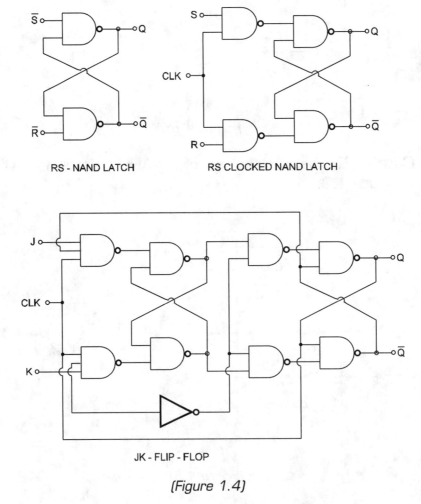

(Figure 1.4)

Symbols

To represent the logic functions in a circuit, logic symbols are used. Unfortunately these symbols are not the same in all parts of the world. The differences found in symbols adopted to represent the basic gates can be observed in Figure 1.5.

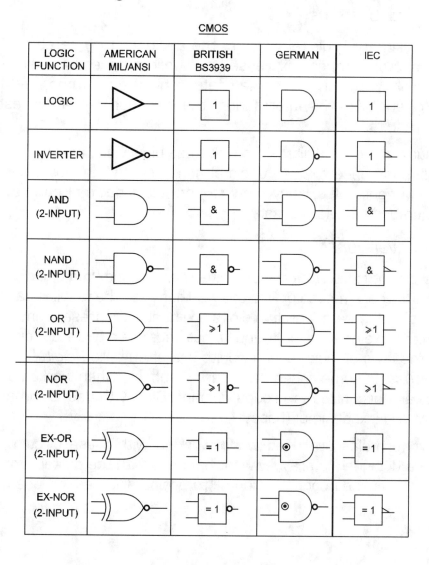

[Figure 1.5]

The symbols used in this book are the American (MIL/ANSI) standards. For the inputs of the logic gates, A, B, C, D, etc., are used, and for the outputs, X, Y, and Z are used. The number that follows the letter, when found, indicates the number of the gate in the IC when the same device contains more than one function.

Numbering Systems

When working with digital circuits, the designer may be able to make conversions of numbers expressed in different bases. A brief explanation about how they work, some useful tables, algorithms, and formulas are given. With this information, the designer can easily reach the desired results.

As mentioned earlier, digital electronics use binary and hexadecimal numbers. Humans generally use decimal numbers. When working with circuits it is important to know how to convert a decimal number into a binary number and vice versa.

Base Conversions

A digital circuit follows the rules of binary arithmetic. In a binary numbering system where a base of 2 is used rather than the familiar base of 10 of the decimal system, only two digits are need to represent any quantity. These digits are designated 1 and 0, but because a two-state circuit also follows the laws of logic, such circuits are often called *digital circuits* or *logic circuits*. In some cases the binary numbers, when worked by digital circuits, are organized in groups forming Binary Coded Decimal (BCD) numbers or Hexadecimal (base of 16).

The next formulas are used when working with these circuits, which allows the designer to preview what happens with the digital information inside a determined configuration or to project the desired one.

Binary to Decimal Conversion

Digital circuits use a base of 2. This means that only two digits are used to represent any quantity. Converting a pure digital number to an equivalent decimal (base of 10) is made using the following formula

Formula

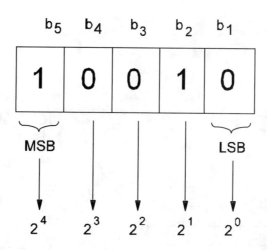

(Figure 1.6)

Binary Pure to BCD conversion:

$$Dn = b1 x 2^0 + b2 x 2^1 + b3 x 2^2 + b4 x 2^3 + ...bn x 2^{n-1}$$

Where:

Dn is the decimal number
b1 is the least significant bit (LSB) of the binary number
b2 to bn -1 are the intermediate bits of the binary number
bn is the most significant digit (MSB) of the binary number
2^0 to 2^n are power of two (see table)

Application Example

Convert the pure binary number 1010100 to decimal:

Apply formula (LSB = 0 and MSB = 1)

$$Dn = 0x2^0 + 0x2^1 + 1x2^2 + 0x2^3 + 1x2^4 + 0x2^5 + 1x2^6$$
$$Dn = 0x1 + 0x2 + 1x4 + 0x8 + 1x16 + 0x32 + 1x64$$
$$Dn = 4 + 16 + 64$$
$$Dn = 80$$

Table 1

Powers of Two

Power of Two	Decimal
2^0	1
2^1	2
2^2	4
2^3	8
2^4	16
2^5	32
2^6	64
2^7	128
2^8	256
2^9	512
2^{10}	1 024
2^{11}	2 048
2^{12}	4 096
2^{13}	8 192
2^{14}	16 384
2^{15}	32 768
2^{16}	65 536
2^{17}	131 072
2^{18}	262 144
2^{19}	524 288
2^{20}	1 048 576
2^{21}	2 097 152
2^{22}	4 194 304
2^{23}	8 388 608
2^{24}	16 777 216
2^{25}	33 554 432
2^{26}	67 108 864
2^{27}	134 217 728
2^{28}	268 435 456
2^{29}	536 870 912
2^{30}	1 073 741 824
2^{31}	2 147 483 648
2^{32}	4 294 967 296

CMOS Basics

Byte to Decimal Conversion

The byte is an 8-bit binary number. The next formula can be used to convert a byte into a decimal:

Formula

Byte to decimal:

$$Dn = b1x2^0 + b2x^1 + b3x2^2 + b4x2^3 + b5x2^4 + b6x2^5 + b7x2^6 + b8x2^7$$

Where:

Dn is the decimal number
b1 to b8 are the bits of the byte
b1 is the MSB (most significant bit)
b8 is the LSB (least significant bit)

Table 2

Decimal Integers to Pure Binaries

Decimal Integer	Binary
00	00000000
01	00000001
02	00000010
03	00000011
04	00000100
05	00000101
06	00000110
07	00000111
08	00001000
09	00001001
10	00001010
11	00001011
12	00001100
13	00001101
14	00001110
15	00001111
16	00010000
17	00010001
18	00010010
19	00010011
20	00010100
21	00010101
22	00010110
23	00010111
24	00011000
25	00011001
26	00011010
27	00011011
28	00011100
29	00011101
30	00011110
31	00011111
32	00100000
33	00100001
34	00100010
35	00100011
36	00100100
37	00100101
38	00100110
39	00100111
40	00101000
41	00101001
42	00101010
43	00101011
44	00101100
45	00101101
46	00101110
47	00100111
48	00110000
49	00110001
50	00110010
51	00110011
52	00110100
53	00110101
54	00110110
55	00110111
56	00111000
57	00111001
58	00111010
59	00111011
60	00111100
61	00111101
62	00111110
63	00111111
64	01000000
65	01000001
66	01000010
67	01000011
68	01000100
69	01000101
70	01000110

BCD to Decimal

A binary-coded decimal is a form of binary representation used in digital electronics where groups of 4 bits represent a decimal digit as shown in Figure 1.7. Conversion to decimal is made as follows:

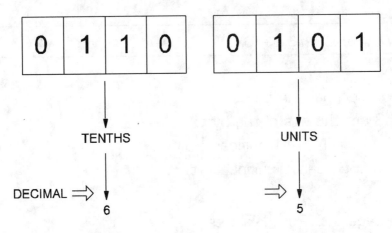

(Figure 1.7)

Formula

BCD to Decimal conversion:

$$Dd = b1 x 2^0 + b2 x 2^1 + b3 x 2^2 + b4 x 2^3$$
$$\text{and}$$
$$Dd < 10$$

Where:

  Dd is the decimal digit
  b1 to b4 are the BCD digits or bits
  b1 is the MSB (most significant bit)
  b4 is the LSB (least significant bit)

CMOS Basics

Application Example

Convert to decimal the BCD 1001 1100.

Solving:

a) Calculating the units decimal digit:

$$Ddu = 0x2^0 + 0x2^1 + 1x2^2 + 0x2^3$$
$$Ddu = 0 + 0 + 8 + 0$$
$$Ddu = 8$$

b) Calculating the tenths decimal digit:

$$Ddt = 1x2^0 + 0x2^1 + 0x2^2 + 1x2^3$$
$$Ddt = 1 + 8$$
$$Ddt = 9$$

The decimal number is 98.

Table 3
Negative Powers of Two

Negative Power of Two	Decimal
2^0	1
2^{-1}	0.5
2^{-2}	0.25
2^{-3}	0.125
2^{-4}	0.062 5
2^{-5}	0.031 25
2^{-6}	0.015 625
2^{-7}	0.007 812 5
2^{-8}	0.003 906 25
2^{-9}	0.001 953 125
2^{-10}	0.000 976 562 5
2^{-11}	0.000 488 281 25
2^{-12}	0.000 244 140 625
2^{-13}	0.000 122 070 312 5
2^{-14}	0.000 061 035 156 25
2^{-15}	0.000 030 517 578 125
2^{-16}	0.000 015 258 789 062 5

Hexadecimal to Decimal Conversion

In this numbering system, digits from 0 through 9 are used as well as the letters from A to F. As in the case of binary and decimal numbers, the value of a hexadecimal number depends on its horizontal position. Conversion to decimal is made using the next formula. Table 4 provides the value of each digit in a hexadecimal numbering system.

Formula

Hexadecimal to decimal conversion:

$$Dn = h1 x 16^2 + h2 x 16^1 + h3 x 16^2 + h4 x 16^4$$

Where:

Dn is the decimal number

h1 to h4 are the hexadecimal digits (*)

h1 is the LSB hexadecimal digit

h4 is the MSB hexadecimal digit

(*) See table below for the equivalent decimal digits to hexadecimal letters.

Application Example

Convert to decimal the hexadecimal F5A2.

Data:

h1 = 2
h2 = A (10)
h3 = 5
h4 = F (15)

Applying the formula and consulting table 45 to powers of 16:

$$Dn = 2 x 16^0 + 10 x 16^1 + 5 x 16^2 + 15 x 16^3$$
$$Dn = 2 x 1 + 10 x 16 + 5 x 256 + 15 x 4096$$
$$Dn = 2 + 160 + 1280 + 61440$$
$$Dn = 62882$$

Table 4
Hexadecimal Digits and Decimal Correspondents

Hexadecimal	Decimal
0	0
1	1
2	2
3	3
4	4
5	5
6	6
7	7
8	8
9	9
A	10
B	11
C	12
D	13
E	14
F	15

Table 5
Powers of 16

Power of 16	Decimal
16^0	1
16^1	16
16^2	256
16^3	4096
16^4	65 536
16^5	1 048 576
16^6	16 777 216

There isn't a formula to make this conversion. To convert a decimal number to pure binary, we have to use an algorithm consisting of successive divisions of the decimal number by the binary base (2). Using this algorithm is explained in the next lines.

Algorithm
Converting decimal to binary:

$$b1.......bn = \left[\frac{dn}{2}\right]\tilde{R}$$

The binary number is found by writing, in the inverse order, the rest of the successive division of the decimal number by two, beginning with the result of the last division.

Where:

b1 to b1 is the binary number

dn is the decimal number

$\bar{R}$ is the result of the divisions in inverse order

Application Example

Convert the decimal number 26 to binary:

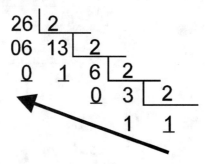

Write in the inverse order: 11010.

Decimal to Hexadecimal

There is not a formula for the conversion of decimal numbers into hexadecimal, but there is an algorithm that can accomplish this.

The hexadecimal number is found by writing, in the inverse order, the rest of the successive division of the decimal number by 16, beginning with the result of the last division.

$$b1.......bn = \left[\frac{dn}{2}\right]\bar{R}$$

Where:

b1 to b1 is the hexadecimal number

dn is the decimal number

$\bar{R}$ is the result of the divisions in inverse order

CMOS Basics

Example:

Convert the decimal number 32452_{10} to hexadecimal. (R is the remainder.) Remember that the algorithms are: 0, 1, 2, 3, 4, 5, 6, 7, 8, 9, A=10, B=11, C=12, D=13, E=14, F=15

$$32452/16 = 2028 \quad\quad R = 14 = E$$
$$2028/16 = 126 \quad\quad R = 12 = C$$
$$126/16 = 7 \quad\quad R = 14 = E$$
$$7/16 = 0 \quad\quad R = 7$$

The result is read upward to give the answer: $7ECE_{16}$.

Logic Functions

The logic functions are blocks used in digital circuits to perform a series of "yes-no" decisions based on the presence or absence of signals at various inputs. The input signals are other "yes and no" signals; and, depending on how the blocks are interconnected, it is possible to build a simple timer through a complex computer.

The following information, including tables, functions, or relationships between the blocks is useful for designers who want work on projects or calculations, or who want to know how a digital block operates in an application.

Combinational Logic

AND Gate

The output of an AND gate is in the high logic level only if either input (A or B) is in the high logic level. Figure 1.8 shows the symbol of an AND gate and the equivalent circuit.

CMOS Sourcebook

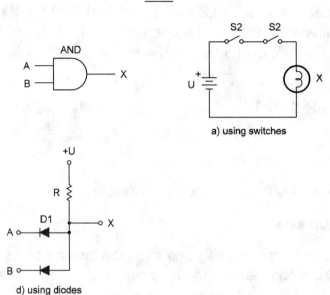

a) using switches

d) using diodes

(Figure 1.8)

Equation

Boolean equation — 2-input AND gate:

$$X = A.B$$

Where:

X is the output logic level

A, B are the input logic levels

Table 6

Truth Table — 2-Input AND Function

A	B	X
0	0	0
0	1	0
1	0	0
1	1	1

CMOS Basics

Equation

Boolean equation — 3-input AND gate

$$X = ABC$$

Where:

X is the output logic level
A, B, and C are input logic levels

Table 7
Truth Table — 3-Input NAND Gate

A	B	C	X
0	0	0	0
0	0	1	0
0	1	0	0
0	1	1	0
1	0	0	0
1	0	1	0
1	1	0	0
1	1	1	1

NAND Gate

The output is high only if the inputs (A and B) are not in the high level. The symbol and the equivalent electric circuit are shown in Figure 1.9.

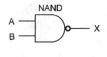

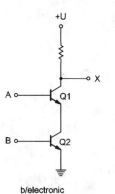

b/electronic

CMOS

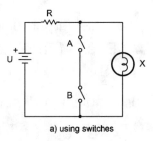

a) using switches

(Figure 1.9)

Equation

Boolean equation — 2-input NAND gate:

$$X = \overline{AB} = \overline{A} + \overline{B}$$

Where:

X is the output logic level

A and B are the input logic levels

Table 8

Truth Table — 2-Input NAND Gate

A	B	X
0	0	1
0	1	1
1	0	1
1	1	0

Equation

Boolean equation — 3-input NAND gate:

$$X = \overline{ABC} = \overline{A} + \overline{B} + \overline{C}$$

Where:

X is the output logic level

A, B, and C are the input logic levels

Table 9

Truth Table — 3-Input NAND Gate

A	B	C	X
0	0	0	1
0	0	1	1
0	1	0	1
0	1	1	1
1	0	0	1
1	0	1	1
1	1	0	1
1	1	1	0

CMOS Basics

OR Gate

The output of an OR gate is high if either input (A or B) or both inputs are in the high logic level. Figure 1.10 shows the symbol and the equivalent electric circuit.

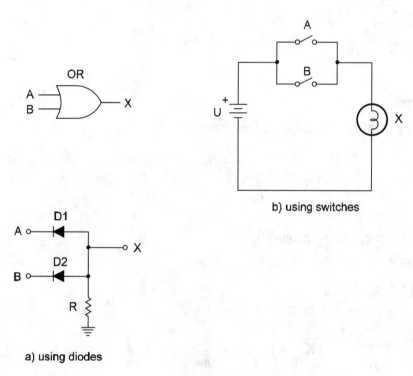

a) using diodes

b) using switches

(Figure 1.10)

Equation

Boolean equation – 2-input OR gate:

$$X = A + B$$

Where:

   X is the output logic level
   A and B are the input logic levels

Table 10
Truth Table — 2-Input OR Gate

A	B	X
0	0	0
0	1	1
1	0	1
1	1	1

Equation
Boolean Equation — 3-input OR gate:

$$X = A + B + C$$

Where:

X is the output logic level

A, B and C are the input logic levels

Table 11
Truth Table — 3-Input OR Gate

A	B	C	X
0	0	0	0
0	0	1	1
0	1	0	1
0	1	1	1
1	0	0	1
1	0	1	1
1	1	0	1
1	1	1	1

CMOS Basics

NOR Gate

The output of a NOR gate is at the high logic level if both inputs (A and B) are in the high logic level. Figure 1.11 shows the symbol and the equivalent circuit of a NOR gate.

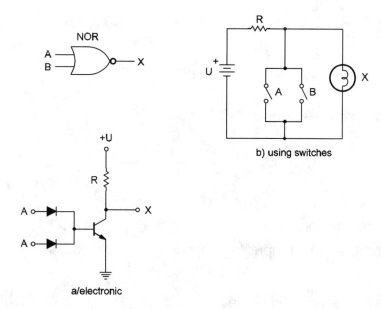

(Figure 1.11)

Equation

Boolean equation — 2-input NOR gate:

$$X = \overline{A+B}$$

Where:
  X is the output logic level
  A and B are the input logic level

Table 12
Truth Table — 2-Input NOR Gate

A	B	X
0	0	1
0	1	0
1	0	0
1	1	0

Equation
Boolean equation — 3-input NOR gate:

$$X = \overline{A + B + C}$$

Where:

X is the output logic level

A, B, and C are the input logic level

Table 13
Truth Table — 3-Input NOR Gate

A	B	C	X
0	0	0	1
0	0	1	0
0	1	0	0
0	1	1	0
1	0	0	0
1	0	1	0
1	1	0	0
1	1	1	0

CMOS Basics

Exclusive-OR Gate

In the output of an exclusive-OR gate, the logic level is high if either input (A or B) is in the high logic level, but not both. The symbol and the equivalent circuit of an exclusive-OR gate are shown in Figure 1.12.

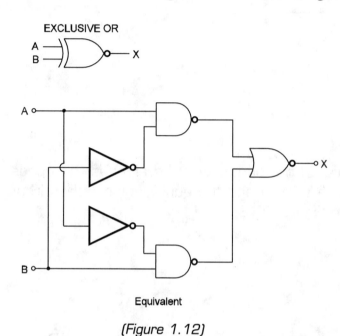

(Figure 1.12)

Equation

Boolean equation — exclusive-OR gate:

$$X = A\overline{B} + \overline{A}B$$

Where:
  X is the output logic level
  A and B are the input logic level

Table 14

Truth Table — Exclusive-OR Gate

A	B	X
0	0	0
0	1	1
1	0	1
1	1	0

Inverter

The logic level at the output of an inverter is the opposite of the input logic level. The symbol and the equivalent electric circuit of an inverter are shown in Figure 1.13.

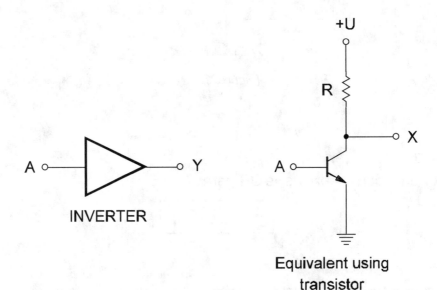

INVERTER

Equivalent using transistor

(Figure 1.13)

CMOS Basics

Equation

Inverter:

$$X = \overline{A}$$

Where:

X is the output logic level

A is the input logic level

Binary Addition

The next general rule is valid when performing binary addition:

Rule

Binary addition:

```
0 + 0 = 0
0 + 1 = 1
1 + 0 = 1
1 + 1 = 1 and 1 to carry
```

Binary Subtraction

The next general rule is valid when performing binary subtraction:

Rule

Binary subtraction:

```
0 - 0 = 0
0 - 1 = 1 and 1 to borrow
1 - 0 = 0
1 - 1 = 0
```

Binary Multiplication

The next rule is valid when performing binary multiplication:

Rule

Binary multiplication:

```
0 x 0 = 0
0 x 1 = 0
1 x 0 = 0
1 x 1 = 1
```

Binary Division

The next rule is valid when performing binary division:

Rule

Binary division:

```
0/0 = ?
0/1 = 0
1/0 = ?
1/1 = 1
```

The Postulates of Boolean Algebra

Laws of Tautology

Repetition by addition or multiplication does not alter the truth value of an element. Figure 1.14 shows the circuit diagram that corresponds to this law.

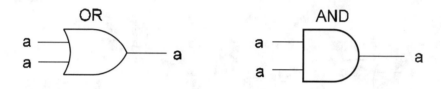

(Figure 1.14)

Law
Tautology:

$$a + a = a$$
$$a \times a = a$$

Mathematic notation (Set Theory):

$$a \cup a = a$$
$$a \cap a = a$$

Logic notation:

$$a \vee a = a$$
$$a \wedge a = a$$

CMOS Sourcebook

Laws of Commutation

Conjunction and disjunction are not affected by sequential change. Figure 1.15 shows the circuit diagrams that correspond to this law.

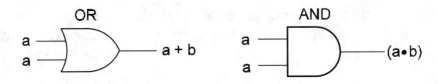

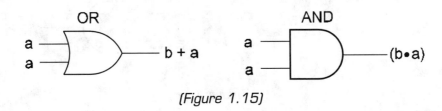

(Figure 1.15)

Law
Commutation:

$$a + b = b + a$$
$$axb = bxa$$

Mathematic notation:

$$a \cup b = b \cup a$$
$$a \cap b = b \cap a$$

Logic notation:

$$a \vee b = b \vee a$$
$$a \wedge b = b \wedge a$$

CMOS Basics

Laws of Association

Grouping does not affect the disjunction or conjunction. Figure 1.16 shows the equivalent circuit diagrams.

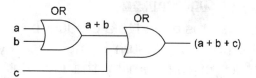

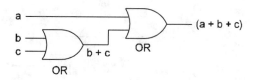

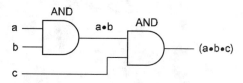

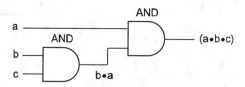

[Figure 1.16]

Laws

Association Laws:

$$a + (b + c) = (a + b) + c$$
$$a x (b x c) = (a x b) x c$$

Mathematic notation:

$$a \cup (b \cup c) = (a \cup b) \cup c$$
$$a \cap (b \cap c) = (a \cap b) \cap c$$

Logic notation:

$$a \vee (b \vee c) = (a \vee b) \vee c$$
$$a \wedge (b \wedge c) = (a \wedge b) \wedge c$$

CMOS Sourcebook

Laws of Distribution

An element is added to a product by adding the element to each member of the product. A sum is multiplied by an element by multiplying every member of the sum by the element. The circuit diagram that corresponds to these laws is shown in Figure 1.17.

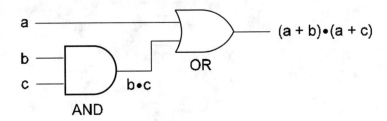

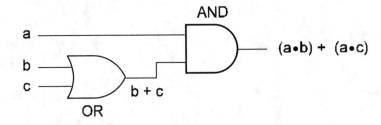

(Figure 1.17)

Laws

Distribution laws:

$$a + (b \times c) = (a + b) \times (a + c)$$
$$a \times (b + c) = (a \times b) + (a \times c)$$

Mathematic notation:

$$a \cup (b \cap c) = (a \cup b) \cap (a \cup c)$$
$$a \cap (b \cup c) = (a \cap b) \cup (a \cap c)$$

Logic notation:

$$a \vee (b \wedge c) = (a \vee b) \wedge (a \vee c)$$
$$a \wedge (b \vee c) = (a \wedge b) \vee (a \wedge c)$$

CMOS Basics

Laws of Absorption

The disjunction of a product by one of its members is equivalent to this member. The conjunction of a sum by one of its members is equivalent to this member. The equivalent logic circuit is shown in Figure 1.18.

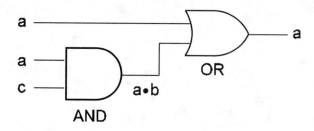

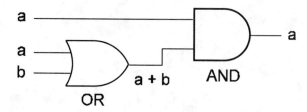

[Figure 1.18]

Laws

Absorption laws:

$a + (a \times b) = a$
$a \times (a + b) = a$

Mathematic notation:

$a \cup (a \cap b) = a$
$a \cap (a \cup b) = a$

Logic notation:

$a \vee (a \wedge b) = a$
$a \wedge (a \vee b) = a$

CMOS Sourcebook

Laws of Universe Class

The sum consisting of an element and the universe class is equivalent to the universe class. The product consisting of an element and the universe class is equivalent to the element. The equivalent circuits are shown in Figure 1.19.

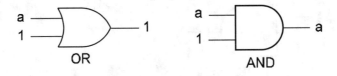

(Figure 1.19)

Laws
Universe Class:

$$a + 1 = 1$$
$$a \times 1 = a$$

Mathematic notation:

$$a \cup 1 = 1$$
$$a \cap 1 = a$$

Logic notation:

$$a \vee 1 = 1$$
$$a \wedge 1 = a$$

CMOS Basics

Laws of Null Class

The sum consisting of an element and the null class is equivalent to the element. The product consisting of an element and the null class is equivalent to the null class. The equivalent logic circuits are shown in Figure 1.20.

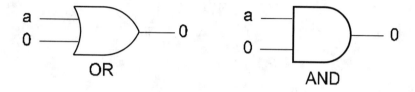

(Figure 1.20)

Laws

Null class:

$$a + 0 = a$$
$$a \times 0 = 0$$

Mathematic notation:

$$a \cup 0 = a$$
$$a \cap 0 = 0$$

Logic notation:

$$a \vee 0 = a$$
$$a \wedge 0 = 0$$

CMOS Sourcebook

Laws of Complementation

The sum consisting of an element and its complement is equivalent to the universe class. The product consisting of an element and its complement is equivalent to the null class. Figure 1.21 shows the logic diagrams that correspond to these laws.

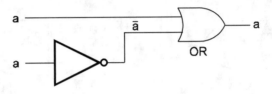

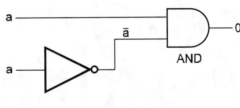

(Figure 1.21)

Laws

Complementation:

$$a + \overline{a} = 1$$
$$a x \overline{a} = 0$$

Mathematic notation:

$$a \cup \overline{a} = 1$$
$$a \cap \overline{a} = 0$$

Logic notation:

$$a \vee \overline{a} = 1$$
$$a \wedge \overline{a} = 0$$

CMOS Basics

Laws of Contraposition

If an element is equivalent to the complement of an element **b**, it is implied that the element **b** is equivalent to the complement of the element **a**. Figure 1.22 shows the equivalent circuit diagram.

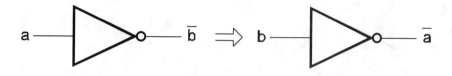

(Figure 1.22)

Laws

Contraposition:

$$a = \overline{b} \Rightarrow b = \overline{a}$$

Mathematic notation:

$$a = b' \geq b = a'$$

Logic notation:

$$a \equiv \approx b. \geq .b \equiv \approx a$$

CMOS Sourcebook

Law of Double Negation

The complement of the negation of an element is equivalent to the element. Figure 1.23 shows the equivalent circuit diagram.

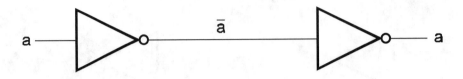

(Figure 1.23)

Laws

Double negation:

$$a = \overline{\overline{a}}$$

Logic notation:

$$a =\sim a'$$

Mathematic notation:

$$a = a'C$$

Laws of Expansion

The disjunction of a product composed of the elements **a** and **b** and a product composed of the element **a** and the complement of the element **b** is equivalent to the element **a**. The conjunction of a sum composed of the elements **a** and **b** and a sum composed of the element **a** and the complement of the element **b** is equivalent to the element **a**. Figure 1.24 shows the circuit diagrams describing these laws.

CMOS Basics

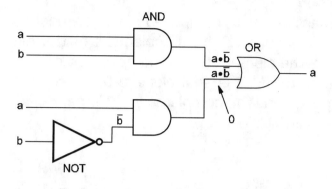

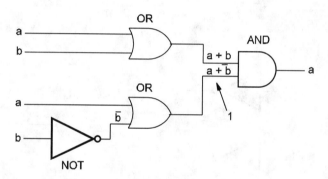

(Figure 1.24)

Laws

Laws of expansion:

$$(a \times b) + (a \times \bar{b}) = a$$
$$(a + b) \times (a + \bar{b}) = a$$

Mathematic notation:

$$(a \cap b) \cup (a \cap b') = a$$
$$(a \cup b) \cap (a \cup b') = a$$

Logic notation:

$$(a \wedge b) \vee (a \wedge \sim b) = a$$
$$(a \vee b) \wedge (a \vee \sim b) = a$$

Laws of Duality

The complement of a sum composed of the elements **a** and **b** is equivalent to the conjunction of the complement of the element **a** and the complement of the element **b**. The complement of a product composed of the elements **a** and **b** is equivalent to the disjunction of the complement of element **a** and the complement of element **a**. Figure 1.25 shows the equivalent circuit diagram.

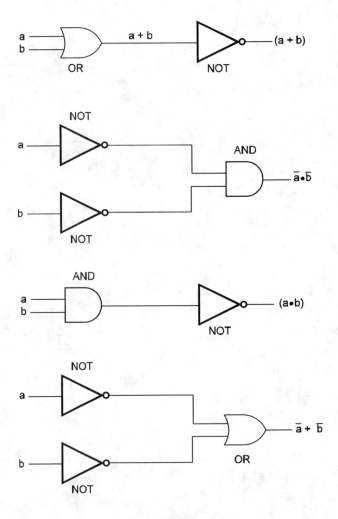

(Figure 1.25)

CMOS Basics

Laws

Laws of Duality:

$$(a+b)' = \overline{a} x \overline{b}$$
$$(axb)' = \overline{a} + \overline{b}$$

Mathematic notation:

$$(a \cup b)' = a' \cap b'$$
$$(a \cap b)' = a' \cup b'$$

Logic notation:

$$\sim(a \vee b) = \sim a \wedge \sim b$$
$$\sim(a \wedge b) = \sim a \vee \sim b$$

Boolean Relationships

Idempoint

Relationship
Addition:

$$a + 0 = a$$
$$a + 1 = 1$$
$$a + a = a$$

Where: $0 \equiv a$

Relationship
Multiplication:

$$0 x a = 0$$
$$1 x a = a$$
$$a x a = a$$

Where: $0 \equiv a$

Commutative

Relationship
Addition:

$$(a+b) = (b+a)$$

Relationship
Multiplication:

$$axb = bxa$$

Associative

Relationship
Addition:
$$(a+b)+c = a+(b+c)$$

Relationship
Multiplication:
$$(axb)xc = ax(bxc)$$

Distributive

Relationship
Distributive:
$$a+(bxc) = ax(b+c)$$
$$a+bxc = (a+b)x(a+c)$$

Absorption

Relationship
Distributive:
$$ax(a+b) = a+axb \equiv a$$

DeMorgan Theorem

Relationship
DeMorgan:
$$\bar{\bar{a}} = a$$
Theorem – 1:
$$\overline{(axb)} = \bar{a}+\bar{b}$$
$$\overline{\overline{(axb)}} = a+b$$
Theorem – 2:
$$\overline{a+b} = \bar{a}x\bar{b}$$
$$\overline{\overline{a+b}} = axb$$

Sequential Logic

Other important functions are found in the CMOS logic ICs.

RS NAND Latch — The RS NAND latch can "remember" logical states even after the controlling input signal(s) have been removed. The circuit is shown in Figure 1.26.

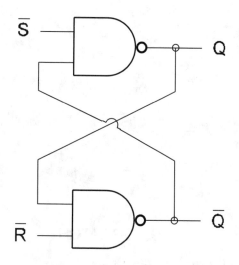

RS - NAND LATCH

(Figure 1.26)

Truth Table:

R	S	Qn+1	/Qn+1
0	0	1	1 (*)
0	1	0	1
1	0	1	0
1	1	Qn	/Qn

(*) Not allowed, no changes

RS NOR Latch — This circuit is an equivalent of the previous latch but uses NOR gates. The circuit is shown in Figure 1.27.

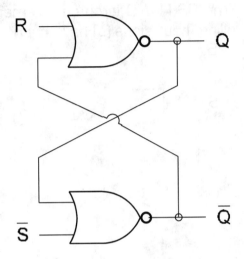

RS NOR LATCH

(Figure 1.27)

Truth Table:

R	S	Qn+1	/Qn+1
0	0	Qn	/Qn
0	1	1	0
1	0	0	1
1	1	0	0 (*)

(*) Not allowed, no changes

CMOS Basics

Clocked RS Nand Latch — When changes in the states of a flip-flop can be synchronized by a clock, the RS NAND Latch can be used in the clocked version shown in Figure 1.28.

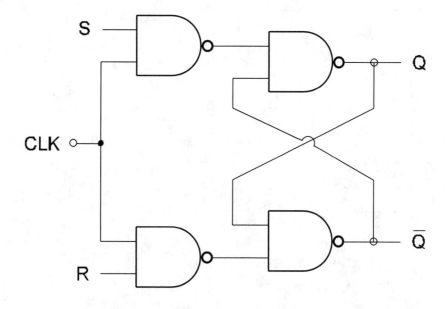

CLOCKED RS NAND SWITCH

(Figure 1.28)

This circuit is sensitive to the clock level. When CLK is 1, the outputs follow the inputs. This means that the inputs change during the time they are active and the outputs will change more than one time. This is not a desirable condition. The problem can be solved with the use of the next configuration.

CMOS Sourcebook

Edge Triggered RS Flip-Flop — The Edge-Triggered RS Flip-Flop or Master-Slave Flip-Flop has the configuration shown in Figure 1.29.

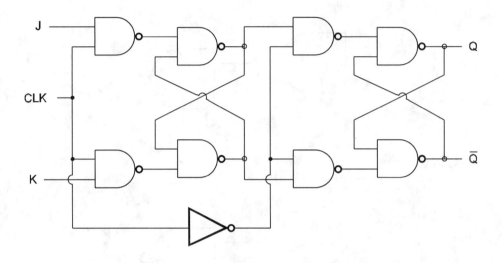

EDGE - TRIGGERED RS FLIP - FLOP

(Figure 1.29)

In this circuit, when CLK=1, only the slave changes its state according to the inputs. Only when the CLK goes to 0 does the condition stored in the slave transfer to the master.

Two parameters are important in this type of Flip-Flop:

- The Hold Time (th) — Defined as the time in which one input may remain stable after the clock transition.
- The Setup Time (ts) — Defined as the time delay in which an input may remain stable before the transition of the clock.

Figure 1.30 illustrates these parameters.

CMOS Basics

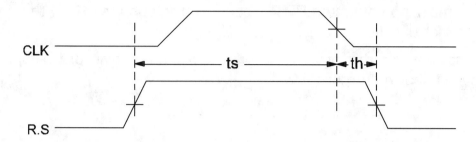

(Figure 1.30)

"Preset" and "Clear" inputs can be added to this type of flip-flop as shown in Figure 1.31.

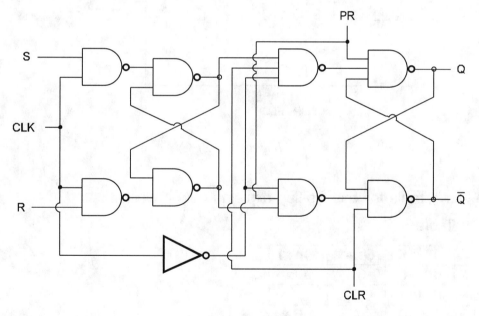

(Figure 1.31)

The J-K Flip-Flop — The flip-flops shown in the last few items don't allow the combination of states R=S=1 because the outputs will go to a state that is not defined.

The solution for this problem is given by another type of flip-flop. The circuit is shown in Figure 1.32.

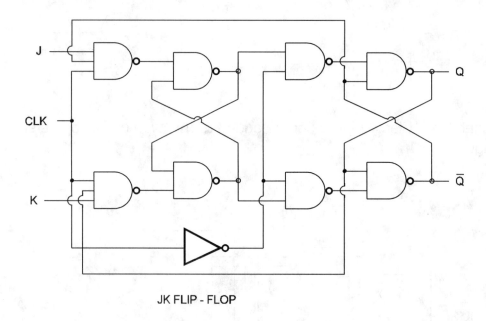

JK FLIP - FLOP

(Figure 1.32)

The Truth Table for this flip-flop explains how it works:

Truth Table:

CLK	J	K	Qn+1	/Qn+1
↓	0	0	Qn	/Qn
↓	0	1	0	1
↓	1	0	1	0
↓	1	1	Qn	/Qn (*)

(*) Toggle
↓ = input transition

Clear (CLR) and Reset (RST) inputs can be added to this circuit as shown in Figure 1.33.

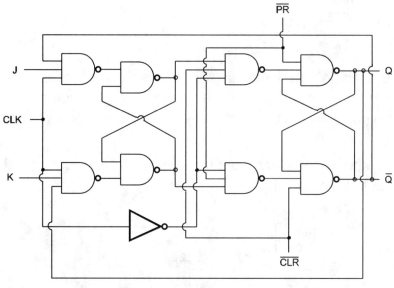

(Figure 1.33)

The D-Latch — This type of flip-flop has only one input named DATA or D as shown in Figure 1.34.

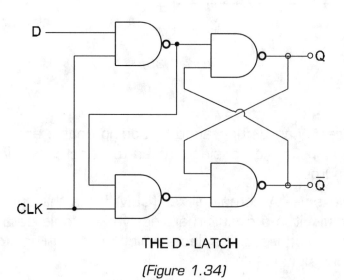

THE D - LATCH

(Figure 1.34)

CMOS Sourcebook

The output depends on this input as shown in the next table:

D	Qn+1
0	0
1	1

The Multiplexer — A multiplexer (MUX) is a circuit that is used to combine two or more digital signals onto a single line by placing the signals at different times. These circuits are also referred to as time-division multiplexers.

Figure 1.35 shows the basic configuration of a MUX that combines two inputs on a single output line.

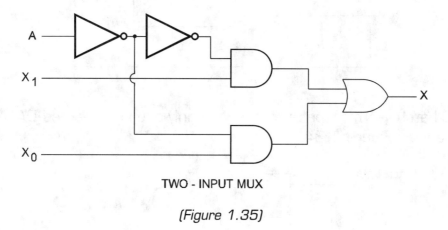

TWO - INPUT MUX

(Figure 1.35)

This simple version can be upgraded by adding enable and clock inputs and addressing inputs. Addressing is used to select which input is connected to the output.

The Demultiplexer — A demultiplexer (DEMUX) is the opposite of the multiplexer circuit. It is a circuit that receives a single data input, and then, based on the address, it selects which of the multiple outputs will receive the input signal.

CMOS Basics

An example is shown in Figure 1.36. Based on the logic level in the A input, it sends the applied signal to the IN input to OUT1 or OUT2.

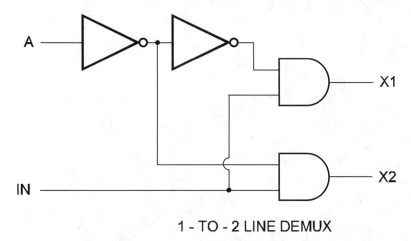

1 - TO - 2 LINE DEMUX

(Figure 1.36)

The Adder — When bits must be added, a special block is used. This block is shown in Figure 1.37 and illustrated in the corresponding Truth Table.

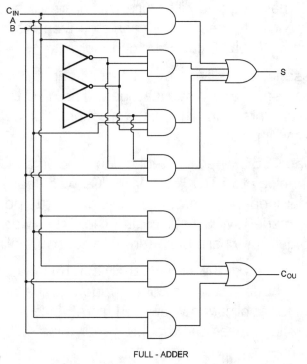

FULL - ADDER

(Figure 1.37)

Truth Table:

INPUTS			OUTPUTS	
A	B	C-in	C-out	S
0	0	0	0	0
0	0	1	0	1
0	1	0	0	1
0	1	1	1	0
1	0	0	0	1
1	0	1	1	0
1	1	0	1	0
1	1	1	1	1

The CMOS Family

Two main logic families are the basis of many parts of modern digital electronic devices: TTL (Transistor-Transistor Logic) and CMOS (Complementary Metal-Oxide Semiconductor).

These logic families have different capabilities and limitations as well as advantages and disadvantages. The TTL technology, created by Texas Instruments, remained the top choice for all designers until a new technology appeared.

It was 1960 when the technology known as CMOS was created as a rival technology to TTL. The CMOS ICs were easy to use, inexpensive, had a high impedance input, and used very little power. Another characteristic that made CMOS better than the TTLs, was its ability to function using voltages anywhere between 3V and 15V, unlike the fixed 5V of the TTLs.

The devices of the CMOS family can be powered with supplies from 3V to 15V. The low current consumption of the devices of this family made it ideal for applications powered from batteries. The original CMOS devices had a low speed, so the TTL family was still ideal for applications where fast operation was needed.

CMOS Basics

The CMOS ICs are found in two versions: 4000A (standard) and 4000B (buffered).

4000A
- Released in 1972
- 3V to 12V operation
- Poor output symmetry
- Outputs sensitive to the input signals
- Current consumption directly proportional to switching frequency

4000B
- Released in 1975
- Uses buffered inverters in a series, increasing linearity (see voltage transfer graph in Figure 1.38)
- Good output symmetry
- Larger propagation delay than series A
- Current consumption directly proportional to switching frequency

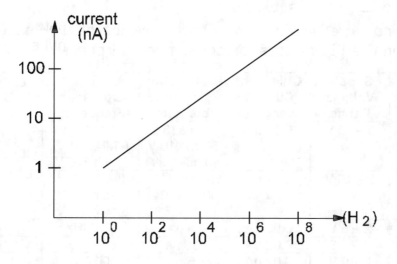

(Figure 1.38)

4000UB

Series 4000UB contains unbuffered circuits, increasing speed. Because their gain is only 23dB, they are ideal for analog applications.

Other Subfamilies:

Many other CMOS subfamilies have been created since 1972 — each one with special characteristics that can be useful for specific applications. Let's consider the 74-series devices.

The idea behind this series is to supply the designer with CMOS devices equivalent in function and packaging of devices of the 7400 TTL family. Many CMOS 74-series subfamilies were created.

- *Standard (74C00)* — This family used normal MOSFET-type CMOS devices. It is now obsolete.
- *High Speed (74HC00)* — Introduced in 1980, it gives the same speed of a regular TTL, but with CMOS characteristics.
- *High Speed (74HCT00)* — Inputs are compatible with TTL outputs.
- *Advanced High Speed (74AC00)* — Typical propagation delays of 5 ns.
- *Advanced High Speed (74HCT00)* — TTL compatible inputs. Propagation delays of 7 ns (typical).

The following table provides the characteristics of the main CMOS families and the TTL devices for comparison purposes:

	Supply Voltage Range	Quiescent Current (per gate)	Propagation delay (per gate)	Maximum Operating Frequency	Fan-out (to LS TTL inputs)
4000B	3 – 15 V	0.01 uA	125 ns (5 V) 50 ns (10 V) 40 ns (15 V)	2 MHz (5 V) 5 MHz (10V) 8 MHz (15 V)	1
4000UB	3 – 15 V	0.01 uA	90 ns (5V) 50 ns (10V) 40 ns (15 V)	3MHz (5 V) 5 MHz (10V) 8 MHz (15V)	1
74HC00	2 – 6 V	0.02 uA	8 ns	40 MHz	10
74HCT00	4.5 – 5.5 V	0.02 uA	10 ns	-	10
74AC00	2 – 6 V	0,02 uA	5 ns	100 MHz	10
74ACT	4.5 – 5.5 V	0.02 uA	7 ns	=	60
TTL LS	4.75 – 5.25	0.5 mA	9 ns	40 MHz	20

CMOS Basics

Characteristics

The characteristics found in the ICs of the CMOS family are the same characteristics as the MOS transistors inside them. Although the CMOS ICs are designed for digital operation, in some cases they can be used as analog amplifiers.

Linear Operation

The voltage transfer characteristic of a CMOS inverter is shown in Figure 1.39. In this figure, you can see that by biasing the input in the linear region of the curve, the device can be used as an analog amplifier.

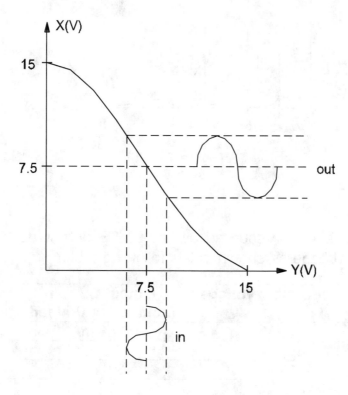

(Figure 1.39)

CMOS Sourcebook

The typical transfer curve of a CMOS inverter, when powered from different sources, is shown by Figure 1.40.

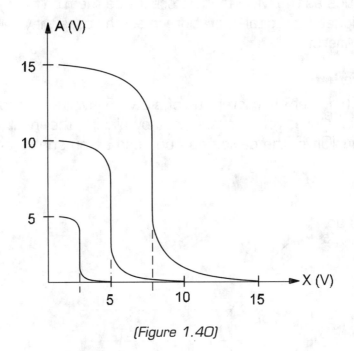

[Figure 1.40]

Digital Operation

The current drained or sourced by any CMOS output depends on the power supply voltage as shown in the following table.

Power Supply Voltage	Output Current (sourced or drained)
5.0 V	0.88 mA
10,0 V	2.2 mA
15 V	8.0 mA

In some ICs, the outputs are not symmetrical and, in these cases, the maximum drained current is different. In this case, the specific type gives the output characteristic of the device.

CMOS Basics

Using CMOS

Electrostatic Protection

The CMOS ICs use an IGFET transistor with a nearly infinite input impedance. This means that if excessive voltage is applied to any input or output of a CMOS circuit, it can break through the gate insulation of the transistors, destroying the device. The human body, under certain conditions, can store electric charges high enough to destroy a CMOS element.

Although the ICs have internal elements to protect against these high voltages, they are limited in their action. Laboratory tests indicate that they can survive spikes reaching high kilovolts. As a precaution, when handling CMOS ICs, never touch the terminals. Also avoid wearing nylon clothing and don't work in places with mats or carpets that tend to store electric charges. It is a good idea to use a grounded wrist strap when working with CMOS ICs.

Unused Inputs

The inputs not used in a project may not "float", or be left free. The input impedance of these inputs can make them sensitive to picking up signal spikes thus affecting the operation of the logic function as it is designed. A second problem with a floating input is that it can put the transistor inside the IC in the linear region of the operation curve where the power consumption is higher. The unused inputs must be tied to ground or Vcc based on the function of the gate. Figure 1.41 shows how the unused inputs are connected.

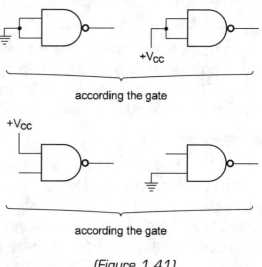

(Figure 1.41)

CMOS Sourcebook

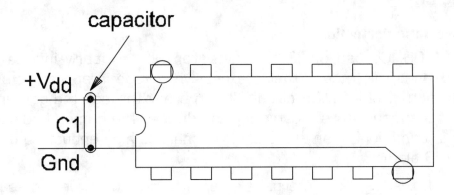

[Figure 1.42]

Decoupling

The fast changes in the logic levels of CMOS ICs cause current spikes that propagate throughout the circuit. If the sensitive elements of the circuit are not decoupled, these spikes can affect their operation, which causes them to go to undetermined states.

When designing a printed circuit board for CMOS applications, the designer must take care with the trails that supply power for the ICs, and, when necessary, must decouple the devices. A ceramic capacitor (0.1 uF) wired between the Vdd and Vss or a CMOS IC is normally enough to decouple the devices, as shown in Figure 1.42.

Power Supplies

CMOS devices have a very low quiescent current making them ideal for applications powered from batteries. Sometimes, however, it is necessary to power them from the AC power line though. Because they are not critical in terms of sourced voltage, in many cases the regulation of power supplies are not needed. Figure 1.43 shows some simple circuits that can be used to power CMOS projects from the AC power line.

CMOS Basics

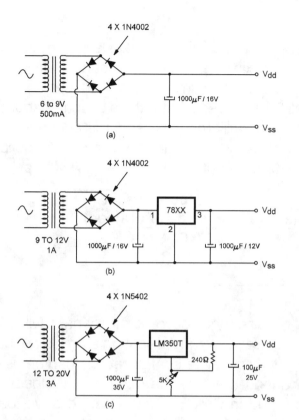

(Figure 1.43)

- This circuit is the simplest and can supply projects not sensitive to voltage changes. The voltage is not regulated and the current depends on the transformer's secondary winding. Transformers with windings ranging from 3V to 9V can be used.
- This is a regulated power supply for currents up to 1A (depending only on the secondary winding of the transformer). The voltage is determined by the IC. The XX determines the voltage. For instance, the 7809 source 9V to the output. The IC must be mounted on a heat sink.
- Finally, for applications where the designer wants to change the voltages, this is the recommended circuit. The output can be adjusted from 1.25V to 15V and the output current is up to 3A. The IC must be mounted on a heat sink.

Interfacing

The CMOS ICs of the 4000 family can't source or drain large amounts of current with their outputs. The maximum current you can get from any device in the family is only enough to drive small loads like LEDs or some other low power device. Typically, the current that can be driven or sourced by any CMOS output remains a low milliamperes, and it is dependent on the power supply voltage. The following table gives the current sourced or drained by the outputs as a function of the power supply voltage.

Power Supply Voltage	Output Current (sourced or drained)
5.0 V	0.88 mA
10.0 V	2.2 mA
15 V	8.0 mA

If you need to drive loads that require more current than what CMOS ICs can source or drain, an intermediate power amplification stage is needed. Depending on the application, many possible configurations can be used. There are low-speed configurations that can be used to turn loads on and off using the logic levels found in CMOS IC outputs. To drive loads with high-speed signals, such as audio and RF, other configurations will be suggested later on.

It is also important to remember to use protection for the driving devices if inductive loads, such as relays, solenoids, or DC motors, are controlled by the stages. The simplest way to protect a device like a bipolar transistor, power FET, or SCR against voltage spikes from fast changes in current is shown in Figure 1.44.

Any general purpose or silicon diode can be used for this task, such as the 1N914, 1N4148, or 1N4002.

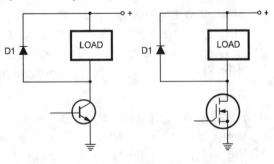

(Figure 1.44)

CMOS Basics

Electromagnetic Interference (EMI)

The fast changes in current across a circuit, especially if it is inductive, can cause electromagnetic interference (EMI). The high-frequency pulses can interfere with devices placed near the circuit, such as radio receivers and other devices that operate using electromagnetic waves. The high voltage spikes generated in this process can also cause the circuit to go into an unstable state (see *decoupling*).

The Circuits

Driving an LED (1)

The simplest configuration designed to drive an LED when the output of the CMOS logic block goes to the high-logic level is shown in Figure 1.45. The LED glows when the output is HIGH and remains off with the output LOW. The resistor R depends on the power supply voltage based on the following table.

Supply Voltage	Resistor R (Ohm)
3 V	100
5 V	330
6 V	470
9 V	680
12 V	1k
15 V	1k2

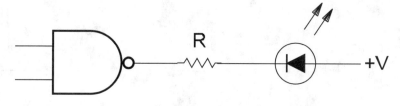

(Figure 1.45)

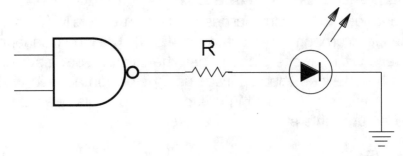

(Figure 1.46)

Driving an LED (2)

The configuration shown in Figure 1.46 is the recommended design to turn on a LED when the output of the CMOS circuit goes to the LOW logic level. When using TTL circuits, the configuration that triggers the load on in the low logic level is preferred because TTL outputs can drain more current than source. In a CMOS device, the drained and sourced current capabilities are the same, so the use of one or another configuration makes no difference. The value of the resistor R also depends on the power supply voltage and whether the table shown in the last circuit is valid.

LEDs in Series

If more than one LED must be driven by the previous circuits, they must be wired in series as shown in Figure 1.47.

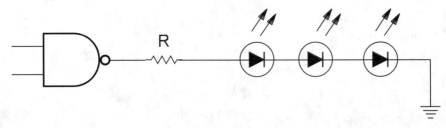

(Figure 1.47)

CMOS Basics

Of course, the power supply voltage and the type of LEDs that are going to be used limit the number of LEDs.

The next table gives the value of R as function of the number of LEDs plugged to the output of a CMOS. We are assuming that the LED is the red type with a voltage fall of about 1.6V. If LEDs of another color are used, the number of units can be reduced because the sum of the voltage fall can't be higher than the power supply voltage.

Power Supply Voltage	2 LEDs	3 LEDs	4 LEDs
5 V	47 ohms	-	-
6 V	100 ohms	-	-
9 V	330 ohms	120 ohms	33 ohms
12 V	560 ohms	330 ohms	180 ohms
15 V	1k	560 ohms	470 ohms

The brightness of the LEDs in each application depends also on the power supply because it determines the maximum output current.

Driving Two LEDs

The circuit shown in Figure 1.48 is a configuration that can switch two LEDs, turning one on when the other is turned off. For instance, the circuit can be used as a logic level indicator with a red and a green LED. The value of the resistor R is chosen according to the table shown in the first application.

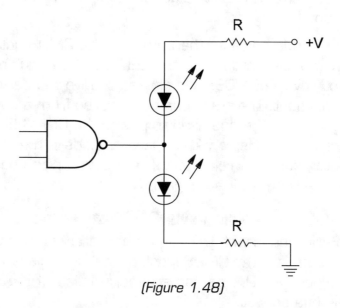

(Figure 1.48)

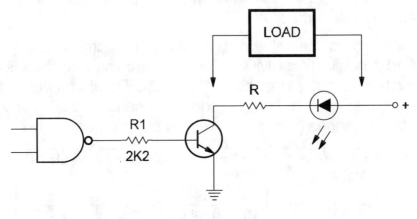

(Figure 1.49)

Low Power Stage Using NPN Transistor

The maximum current sourced or drained by all the circuits shown previous to this is too weak for some applications. The circuit in Figure 1.49 can drive loads up to 100 mA, such as relays, lamps, small DC motors, and other devices with a general-purpose NPN transistor, such as BC548 or any equivalent.

The load is on when the output of the CMOS gate goes to the high logic level. In the application, the LED can be a high-power unit not driven directly by the CMOS. R is chosen to give the desired current through the component. The stage can be powered from a power supply with voltages higher than the one used to power the CMOS blocks. Remember to protect the transistor with a parallel diode if the load is inductive — like a motor, solenoid, relay, etc. The resistor in the base of the transistor (Rx) can be reduced if a low-gain transistor is used.

Low Power Stage Using PNP Transistor

To drive loads with the same characteristics as the last circuit but send the output of the CMOS block to the low logic level, use the circuit shown in Figure 1.50. The same general rules of the last circuit are valid for this one as well.

CMOS Basics

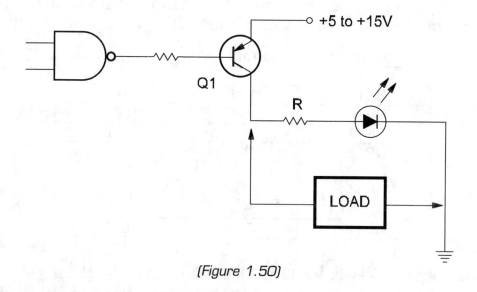

(Figure 1.50)

Medium Power Stage Using Darlington NPN Transistor

If your application needs more than 100 mA to be driven, the block shown in Figure 1.51 can be used. This block uses a NPN Darlington transistor to supply a load with currents up to more than 5 amperes. The load current depends on the transistor used.

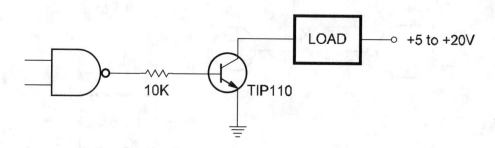

(Figure 1.51)

CMOS Sourcebook

The next table gives some suitable Darlington transistors for this task and the maximum recommended current in the application.

Transistor	Current (Ic)	Voltage (Vce)
TIP110	1.25	60 V
TIP111	1.25	80 V
TIP112	1.25	100 V
TIP120	3 A	60 V
TIP121	3 A	80 V
TIP122	3 A	100 V
TIP140	5 A	60 V
TIP141	5 A	80 V
TIP142	5 A	100 V

Medium Power Stage Using Darlington PNP Transistors

The equivalent configuration using PNP Darlington transistors is shown in Figure 1.52. The transistors are shown in the table. The load is on when the output of the CMOS IC goes to the low logic level. Remember to install the transistor in a heat sink.

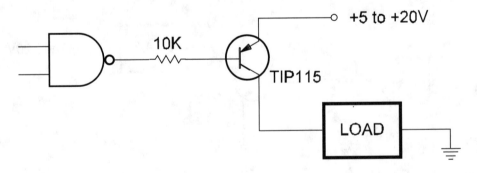

(Figure 1.52)

CMOS Basics

Transistor	Current (Ic)	Voltage (Vce)
TIP115	2 A	60 V
TIP116	2 A	80 V
TIP117	2 A	100 V
TIP125	3 A	60 V
TIP126	3 A	80 V
TIP127	3 A	100 V
TIP145	5 A	60 V
TIP146	5 A	80 V
TIP147	5 A	100 V

Medium Power Stage Using Discrete Darlington NPN Transistors

A discrete bipolar transistor can be wired to form a Darlington stage. The high current gain of this stage is suitable for applications such as driving high-current loads from the low-current output of CMOS circuits. Figure 1.53 shows a stage using two NPN transistors. Q1 is any general purpose NPN transistor, such as the 2N3904 or BC548. Q2 depends on the current needed to drive the load. The next table offers some suggestions for Q2. It is important to remember that the characteristics in the table are the maximum, and in a real circuit it is necessary to take in account the safe operating range. The product of the voltage across the transistor will not exceed the power dissipation any time.

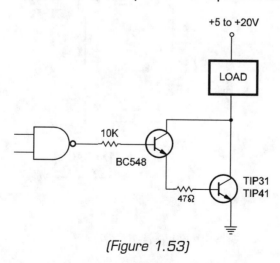

(Figure 1.53)

CMOS Sourcebook

Transistor	Current (Ic)	Voltage (Vce)
BD135	1 A	45 V
BD137	1 A	60 V
BD139	1 A	80 V
TIP31	3 A	40 V
TIP31A	3 A	60 V
TIP31B	3 A	80 V
TIP31C	3 A	100 V
TIP41	6 A	40 V
TIP41A	6 A	60 V
TIP41B	6 A	80 V
TIP41 C	6 A	100 V

Medium Power Stage Using Discrete Darlington PNP Transistors

The equivalent configuration for PNP discrete transistors forming a Darlington stage is given in Figure 1.54. Q1 is any PNP general-purpose silicon transistor, such as the BC558 or equivalent. Q2 is chosen according the current needed by the load. The next table offers some suggestions.

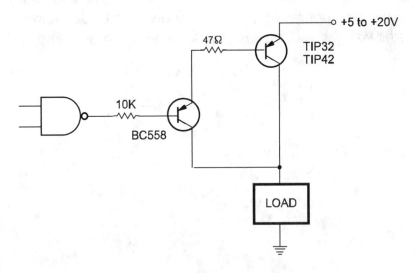

(Figure 1.54)

CMOS Basics

Transistor	Current (Ic)	Voltage (Vce)
BD136	1 A	45 V
BD138	1 A	60 V
BD140	1 A	80 V
TIP32	3 A	40 V
TIP32A	3 A	60 V
TIP32B	3 A	80 V
TIP32C	3 A	100 V
TIP42	6 A	40 V
TIP42A	6 A	60 V
TIP42B	6 A	80 V
TIP42C	6 A	100 V

Medium Power Stage Using Complementary Transistors (I)

Another way to drive high-current loads is by using a discrete bipolar transistor, which is shown in Figure 1.55. In this case, use a directed coupled stage with a complementary transistor. The load is on when the output of the logic block goes to the high logic level.

Q1 is any general-purpose NPN transistor, and Q2 is a PNP transistor that can be chosen from the table in the previous block. The transistor must be mounted on a heat sink. Also, if the load is inductive, a protection diode must be added in parallel.

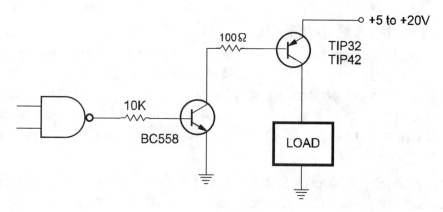

(Figure 1.55)

Medium Power Stage Using Complementary Transistors (II)

To turn on a load when the output of the CMOS block goes to the low logic level, use a circuit such as the one shown in Figure 1.56. This circuit operates the same as the previous circuit. Q1 is any PNP general-purpose transistor, such as BC558. Q2 can be a BD135 or any other device suggested in the table for medium power stage using discrete transistors (Darlington). The same precautions with the heat sink and inductive loads must be observed in this circuit as well.

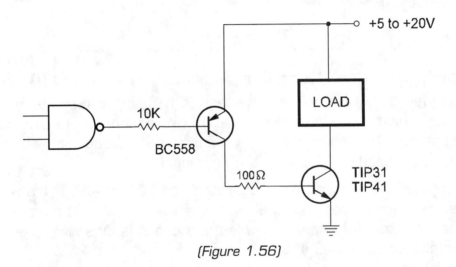

(Figure 1.56)

High Power Using Power FETs

A power MOSFET can be used to drive high-current loads from CMOS outputs by adding only one resistor to the circuit (in some cases even this resistor can be omitted) and plugging the output of the CMOS IC directly into the gate of the transistor. A simple configuration that turns on a load when the output of the CMOS IC goes to the high logic level is shown in Figure 1.57.

Any Power MOSFET (such as the ones of the IRF series) can be used in your application, because they all can control currents, and in some cases currents as high as 100 A.

CMOS Basics

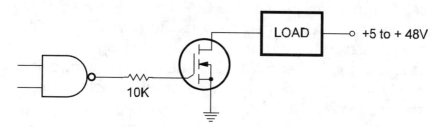

(Figure 1.57)

It is important to observe that the Rds (resistance between drain and source when the transistor is on) have very low values only if the gate voltage is high (above 10 V). This means that in normal condition when low-voltage CMOS blocks are driving this block, the voltage fall across the transistor can be enough to produce a large amount of heat. Consider the use of bipolar transistors if a Power MOSFET can't drive your load as you want.

Observation: the normal Power MOSFET is an N channel unit. If you need to drive a load when the logic level is LOW consider the use of an Inverter stage.

Driving a SCR in a low voltage DC Circuit

Sensitive SCRs, such as the ones in the 106 series, can be triggered by the low power pulses produced in the output of a CMOS circuit. These devices are sensitive enough to be plugged directly into the output of the logic block, as shown in Figure 1.58.

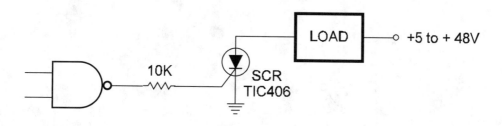

(Figure 1.58)

CMOS Sourcebook

Remember that once triggered, the SCR can't be turned off by pulses applied to the gate — even the negatives. To turn off the SCR in a DC circuit, it is necessary to cut the power-supply voltage or to put a short in the anode with the cathode. The SCR TIC106 can control loads up to 3A. Remember that a voltage fall of about 2V is noted across a SCR when in the conductive state.

SCR/CMOS Flip-Flop

To turn an SCR on and off from CMOS circuits, the flip-flop shown in Figure 1.59 can be used. This circuit can directly trigger loads up to 3A if the SCRs in the 106 series are used. The SCRs must be mounted on heat sinks, and the power supply voltage can typically range from 9V to 40V. Remember that a voltage fall of about 2V is noted across a SCR when in the conductive state.

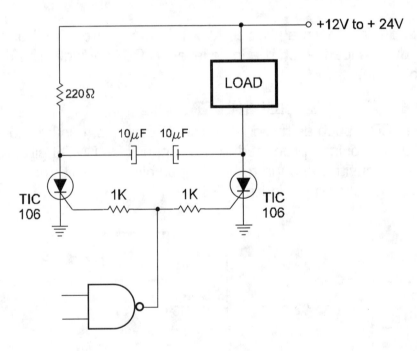

[Figure 1.59]

Driving an SCR in a High-Voltage AC Circuit (Half Wave)

In the AC circuits, sensitive SCRs, such as those from the 106 series, can be triggered directly from the CMOS outputs as shown in the Figure 1.60. In some cases, a resistor is necessary to prevent erratic triggering. For example, when the TIC106 is used, this resistor can range from 10k to 470k.

It is important to make sure that the low-voltage section of the circuit has a common ground with the high-voltage section. Another point to consider is that in AC circuits, the SCR remains on only during the time the CMOS applies voltage to the gate. When the signal is removed, the SCR returns to the low logic level.

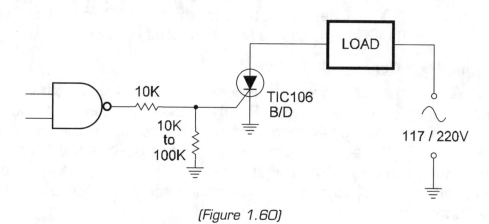

(Figure 1.60)

Driving an SCR in a High-Voltage AC Circuit (Full Wave)

The last driving block is a half-wave circuit, which means that only the positive half cycles of the power-supply voltage are sent to the load. The load operates with about half of the total power. This circuit presents no problems if the load is an incandescent lamp or a heater. If load must be driven with full-wave voltages, the version shown in Figure 1.61 should be used. As in the previous case, the SCR remains on only during the time the output of the CMOS block is in the high logic level.

CMOS Sourcebook

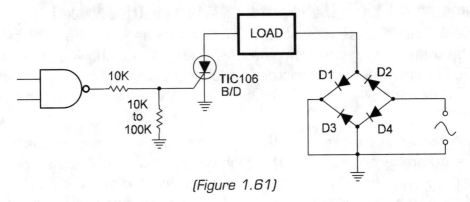

(Figure 1.61)

The diodes are chosen according the current of the load. The rated voltage must be higher than the peak voltage of the power-supply line. For instance, 200V diodes are suitable for the 117VAC power line. Types such as the 1N4004 can be used to drive loads up to 1A.

Driving High Current SCRs (NPN)

The SCRs of the 106 series are very sensitive, needing less than 1 mA to be triggered. If high-power SCRs that need more current to be triggered are used in an application, the best circuit to use is the one shown in Figure 1.62. These circuits are designated as SCRs of the TIC series, which need currents in the range between 10 mA and 100 mA to be triggered. In this application, the SCR is triggered when the output of the CMOS block goes to the high logic level.

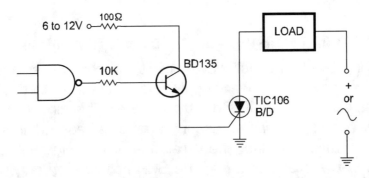

(Figure 1.62)

CMOS Basics

The transistor is a BD135 or any other medium-power silicon NPN transistor and R is between 47 ohms and 220 ohms. The supply voltage for this stage is between 6V and 15V. This configuration can be used either in DC or AC circuits. It is also important to remember that the low-power circuit with the CMOS block has a common ground with the high-power circuit where the SCR is placed.

Driving High-Current SCRs (PNP)

To trigger an SCR on when the output of a CMOS circuit goes to the low logic level, the circuit shown in Figure 1.63 can be used. The transistor is the BD136 or BD138 and the resistor R is between 47 ohms and 220 ohms based on the current needed to trigger the SCR. The voltage is between 6V and 15V.

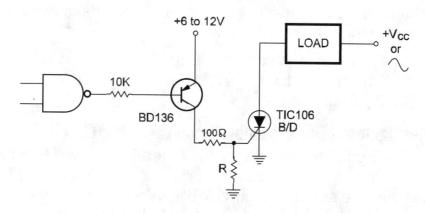

[Figure 1.63]

The resistor R is needed to bias some types of SCRs if they tend to erratically trigger. Values are between 1k and 10k.

Driving a TRIAC

In general, TRIACs need more current than the SCRs to be triggered, and some can't be plugged directly into the output of a CMOS circuit. The use of a transistor to drive these devices is shown by Figure 1.64. The transistor is a BD135, BD137, or any other medium-power NPN tran-

CMOS Sourcebook

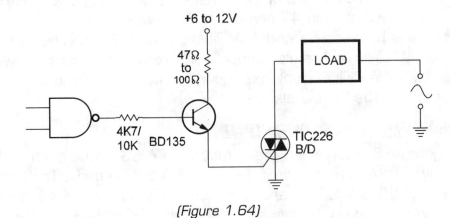

(Figure 1.64)

sistor. The resistor R is between 47 ohms and 220 ohms and is chosen based on the sensitivity of the SCR and the voltage +V. This voltage stays between 9V and 15V.

SCRs of the TIC series, such as the TIC226, can be used in this circuit. Observe that the ground for the AC stage is common with the ground with the CMOS circuits.

Driving an Isolated-Gate Bipolar Transistor (IGBT)

Isolated-gate bipolar transistors or IGBTs are high-power devices used to control high-current loads in DC circuits. To drive an IGBT, use a circuit like the one shown in Figure 1.65. When triggered, IGBTs have the same characteristics found in Power FETs. But, when considering the load circuit, they have the same characteristics of high-power bipolar transistors, too.

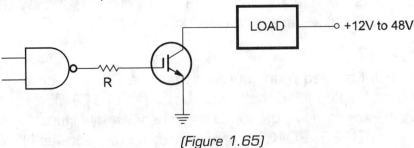

(Figure 1.65)

CMOS Basics

Using an Optocoupler (I)

The previous circuits powered devices from the AC power line and were controlled from CMOS outputs. They have as an important limitation the presence of a common ground for the high-voltage stage and the CMOS circuit. This common ground can represent danger of shock hazards if any part of the low-power circuit is touched.

A solution to trigger high-power devices from the AC power line is the use of optocouplers. These devices use an LED that acts on a sensor and, in this way, isolates the source of signal from the receiver. Isolation voltages of more than 7000V are common in these devices. The first application of an optocoupler driving a transistor is shown in Figure 1.66.

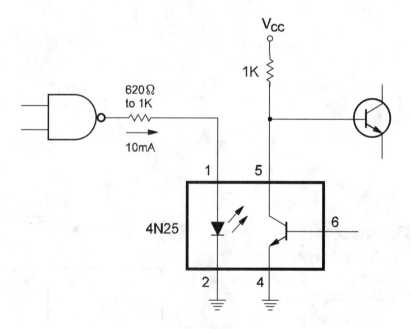

(Figure 1.66)

Transistors can be used to drive an SCR or other high-power circuit. Notice that this circuit is triggered on when the logic level at the output of the CMOS block is low. To drive the circuit when the output of the CMOS output is high, you can use the circuit shown in Figure 1.67.

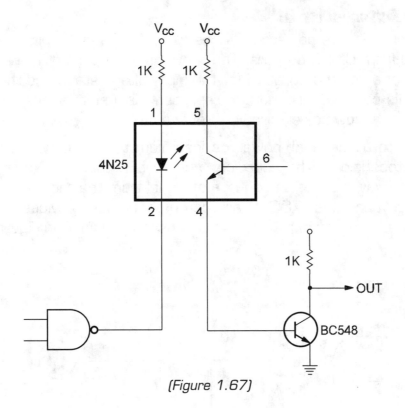

(Figure 1.67)

Using an Optocoupler (II)

Figure 1.68 shows how to use an optocoupler to transfer signals from one to another CMOS stage. This circuit can be used to isolate blocks using CMOS elements.

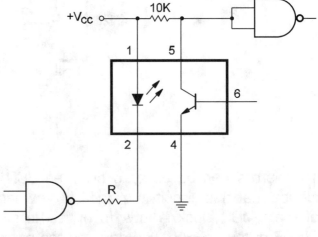

(Figure 1.68)

CMOS Basics

The resistor R depends on the power supply voltage based on the next table:

Supply Voltage	Resistor R
5 V	120 ohms
6 V	150 ohms
9 V	220 ohms
12 V	470 ohms
15 V	680 ohms

NOTE: The values in the table are approximated and depend on the sensitivity of the phototransistor. Depending on the application, a tolerance range of 50 percent can be adopted.

Driving a Triac with an Optodiac (127 VAC)

The optodiac MOC3010 is a device formed by an Infrared LED and an optodiac, a trigger element for Triacs. This device is ideal for triggering Triacs from CMOS (and other logic outputs) with the necessary isolation from one stage to other. Any Triac can be triggered from this circuit and the value of R depends on the power-supply voltage used in the CMOS blocks as shown in the next table. Figure 1.69 shows the device.

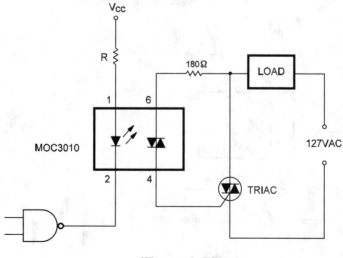

(Figure 1.69)

CMOS Sourcebook

Power Supply Voltage	Resistor R
5 V	180 ohms
6 V	220 ohms
9 V	470 ohms
12 V	820 ohms
15 V	1k2 ohms

It is important to remember that there are two other optocouplers in same family (MOC3011 and MOC3012) that are more sensitive and need less current to be triggered. If the Triac doesn't trigger with the values of the resistors as shown in the table, you can reduce them or use another optodiac.

Driving a Triac with an Optodiac (220/240 VAC)

To trigger loads in a 220V / 240AC power line, the optodiac MOC3020 (and other more sensitive units such as the MOC3021 and MOC3022) can be used in the configuration shown in Figure 1.70. The resistor R is chosen by the same table as given in the previous circuit.

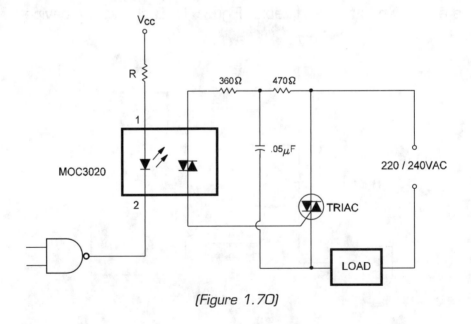

(Figure 1.70)

CMOS Basics

Interfacing

CMOS logic circuits can interface with other logic families, such as TTL, operational amplifiers, and comparators. In some cases, it is not possible to plug the output of one circuit directly to the input of the other. Some kind of interface circuit is needed.

TTL to CMOS (5 V)

The first case is shown in Figure 1.71. A TTL block drives a CMOS block and both are powered from a 5V supply. The 2k2 resistor is used to pull up the output of the TTL gate because the very high input impedance of the CMOS blocks acts as an open circuit.

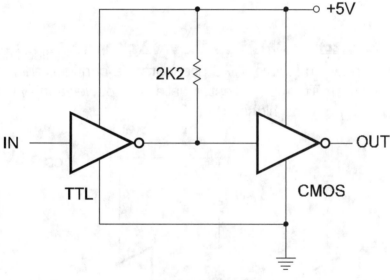

(Figure 1.71)

TTL to CMOS (Different Voltages)

If the CMOS circuit is powered from a voltage higher than 5V, the circuit shown in Figure 1.72 is the recommended for use. The transistor is wired in the common emitter configuration and acts as a "buffer-inverter" for the signals coming from the TTL output. Any general-purpose NPN transistor can replace the 2N2222 in this application.

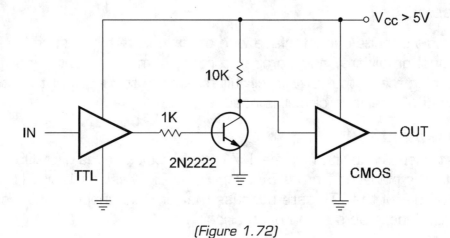

(Figure 1.72)

TTL Open Collector to CMOS (Different Voltages)

The diagram shown in Figure 1.73 presents the configuration to drive a CMOS gate from a TTL open collector gate. The power-supply voltage for the CMOS block must be higher than 5V.

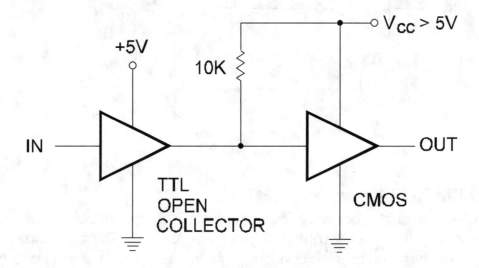

(Figure 1.73)

CMOS Basics

CMOS to TTL (5V)

Figure 1.74 shows how a CMOS circuit can drive a TTL input when both are supplied from a 5V source. The resistor acts as a load for the CMOS output.

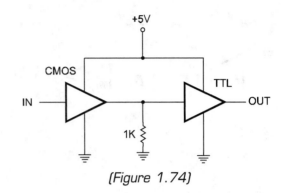

(Figure 1.74)

CMOS to TTL (Different Voltages)

If the CMOS block is powered with a voltage higher than 5V, the correct way to drive a TTL circuit is shown in Figure 1.75. The transistor can be replaced by any general purpose NPN transistor, such as BC548.

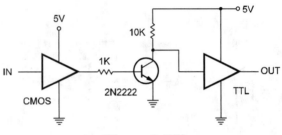

(Figure 1.75)

CMOS to TTL Using a CMOS Buffer

Another way to drive a TTL input from a CMOS output is shown in Figure 1.76. One buffer-inverter of the 6 existing in a 4049 IC is used in this configuration. The buffer-inverter is powered from the same 5V supply that powers the TTL gate, and the CMOS block must be powered from a voltage source higher than 5V.

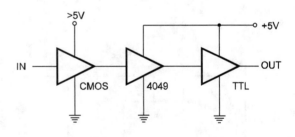

(Figure 1.76)

Operational Amplifier to CMOS (Same Voltage)

Operational amplifiers of voltage comparators can drive a CMOS stage using only a 10k ohm resistor, as shown in Figure 1.77. This configuration is valid if the voltage sources are the same for the two blocks.

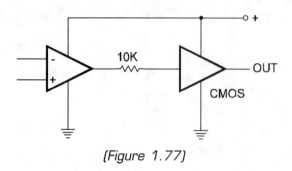

(Figure 1.77)

Operational Amplifier to CMOS (Different Voltages)

The circuit shown in Figure 1.78 is used to drive a CMOS input from an operational amplifier or a voltage comparator and they are powered with different voltages. The diodes are used to protect the CMOS inputs. Any general-purpose silicon diode, such as the 1N914 or 1N4148, can be used in this task.

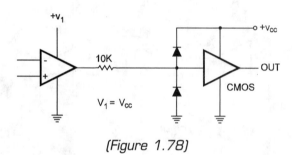

(Figure 1.78)

TTL to CMOS Using Optocoupler

Our last circuit is suggested for use to drive CMOS inputs from TTL outputs using an opto-coupler. The configuration shown in Figure 1.79 has a principal advantage of isolation between the circuits. The optocoupler can be a 4N25 or any equivalent.

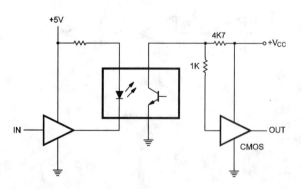

(Figure 1.79)

SECTION 2
FUNCTIONAL DIAGRAMS AND INFORMATION FOR DESIGNERS

Functional Diagrams and Information for Designers

The following pages provide basic information about the most important devices of the 4000 CMOS Series. The information is comprised, in order of appearance, of the following:

- Device Name – This is the number 4XXX. The letters and numbers identifying the manufacturer are omitted.
- Description – This tells what the device is, the number of independent circuits, and what it makes.
- Functional diagram or/and package – This is the functional diagram of the device using logic symbols, and in some cases, how basic external components are plugged. Also the pinout is shown, which is important information for any new project.
- Pin Names – These are the meaning of each abbreviation used to design the pins of the device.
- Truth Table or Waveform – If a component has a logic table to represent operation states, it will be given in this item.
- When there is no truth table, the logic waveforms – or how the signals appear in the output as function of an input signal – appears.
- Operation Mode (How to Use) – Basic information about the use and application of the device, as well as logic levels that can be applied to the input during normal operation is provided in this item.
- Electrical Characteristics – Voltages, currents, and frequencies are important information for the designers. This item lists only the most important of the characteristics. The values are a typical average because they can change from manufacturer to manufacturer. With averaged values the reader can have an idea if the device can or cannot be used in a determined application.
- Other Devices (with the Same Function) – If any device exists with the same function and only small differences in the characteristics or that can be used in similar applications, they will be listed here.
- Applications – Common applications for the devices. In Part 3 we will give details in some circuits where the devices are used.
- Observations – Any final observations about the device will be included in this item.

DEVICES

Ordered by Number

4000:	Dual 2-Input NOR gate plus inverter
4001:	Quad 2-Input NOR gate
4002:	Dual 4-Input NOR gate
4006:	18-Stage Shift Register
4007:	Dual Complementary Pair Plus Inverter
4008:	4-Bit Full Adder
4009:	Hex Buffer (Inverting)
4010:	Hex Buffer (Non-Inverting)
4011:	Dual 2-Input NAND Gate
4012:	Dual 4-Input NAND Gate
4013:	Dual D Flip-Flop
4014:	8-Stage Static Shift Register
4015:	Dual 4-Bit Static Shift Register
4016:	Quad Bilateral Switch
4017:	Decade Counter/Divider with 10 Decoded Outputs
4018:	Presettable Divide-by-N Counter
4019:	Quand AND-OR Select Gate
4020:	14-Stage Ripple-Carry Binary Counter/Divider
4021:	8-Stage Static Shift Register
4022:	Divide-by-8 Counter
4023:	Triple 3-Input Buffered NAND Gate
4024:	7-Stage Ripple-Carry Binary Counter/Divider
4025:	Triple 3-Input Buffered NOR Gate
4027:	Dual J-K Master/Slave Flip-Flop with Set and Reset
4028:	BCD-to-Decimal Decoder
4029:	Pre-settable Binary/Decade Up/Down Counter
4030:	Quad Exclusive-OR Gate
4031:	64-Stage Static Shift Register
4034:	8-Stage Tri-State Bi-directional Parallel/Serial Input/Output Bus Register
4035:	4-Bit Parallel In/Parallel Out Shift Register
4041:	Quad True/Complement Buffer
4042:	Quad Clocked D Latch
4043:	Tri-State NOR R/S Latches
4044:	Tri-State NAND R/S Latches
4046:	Micro power Phase-locked-Loop
4047:	Low Power Monostable/Astable Multivibrator
4048:	Tri-State Expandable 8-Function 8-Input Gate
4049:	Hex Inverting Buffer
4050:	Hex Non-Inverting Buffer
4051:	Single 8-Channel Analog Multiplexers/Demultiplexers
4052:	Dual 4-Channel Analog Multiplexers/Demultiplexers
4053:	Triple 2-Channel Analog Multiplexers/Demultiplexers
4066:	Quad Bilateral Switch
4069:	Inverter Circuits
4070:	Quad 2-Input Exclusive-OR Gate
4071:	Quad 2-Input OR Buffered B Series Gate
4072:	Dual 4-Input OR Buffered B Series Gate
4073:	Double Buffered Triple 3-Input AND Gate
4075:	Double Buffered Triple 3-Input OR Gate
4076:	Tri-State Quad D Flip-Flop
4081:	Quad 2-Input AND Buffered B Series Gate
4082:	Dual 4-Input AND Buffered B Series Gate
4089:	BCD Rate Multiplier
4093:	Quad 2-Input NAND Schmitt Trigger
4094:	8-Bit Shift Register/Latch with Tri-State Outputs
4099:	8-Bit Addressable Latches
40106:	Hex Schmitt Trigger
40160:	Decade Counter with Asynchronous Clear

Functional Diagrams and Information for Designers

40161: Binary Counter with Asynchronous Clear
40162: Decade Counter with Synchronous Clear
40163: Binary Counter with Synchronous Clear
40174: Hex D Flip-Flop
40175: Quad D Flip-Flop
40192: Synchronous 4-Bit Up/Down Decade Counter
40193: Synchronous 4-Bit Up/Down Binary Counter
4503: Hex Non-Inverting Tri-State Buffer
4510: BCD Up/Down Counter
4511: BCD-to-7 Segment Latch Decoder/Driver
4512: 8-Channel Buffered Data Selector
4514: 4-Bit Latched 4-to-16 Line Decoders
4515: 4-bit Latched 4-to-16 Line Decoders
4516: Binary Up/Down Counter
4518: Dual Synchronous Up Counter
4520: Dual Synchronous Up Counter
4522: Programmable Divide-by-N 4-Bit Binary Counter
4526: Programmable Divide-by-N 4-Bit Binary Counter
4528: Dual Monostable Multivibrator
4529: Dual 4-Channel or Single 8-Channel Analog Data Selector
4538: Dual Monostable Multivibrator
4541: Programmable Timer with Oscillator
4543: BCD-to-7 Segment Latch/Decoder/Driver for Liquid Crystal Displays
4723: Dual 4-Bit Addressable Latch
4724: 8-Bit Addressable Latch

Ordered by Function

NAND Gates
4011: Quad 2-Input
4012: Dual 4-Input
4023: Triple 3-Input
4093: Quad 2-Input Schmitt Trigger

NOR Gates
4001: Quad 2-Input
4002: Dual 4-Input
4025: Triple 3-Input

AND Gates
4073: Triple 3-Input
4081: Quad 2-Input
4082: Dual 4-Input

OR Gates
4071: Quad 2-input
4072: Dual 4-Input

Complex Gates
4030: Quad Exclusive-OR
4048: Tri-State Expandable 8-Function 8-Input Gate
4070: Quad Exclusive-OR

Inverters/Buffers
4007: Dual Complementary Pair Plus Inverter
4009: Hex Inverting Buffer
4010: Hex Non-Inverting Buffer
4049: Hex Inverting Buffer
4050: Hex Non-inverting Buffer
4069: Hex Inverter
40106: Hex Schmitt Trigger
4503: Hex 3-State Buffer

Decoders/Encoders
4028: BCD-to-Decimal/Binary-to-Octal Decoder
4511: BCD-to-7 Segment Latch/Decoder/Driver
4514: 4-Bit Transparent Latch 4-to-16 Line Decoder (High)
4515: 4-Bit Transparent Latch 4-to-16 Line Decoder (Low)

Mux/Demux/Bilateral Switches
4016: Quad Analog Switch/Multiplexer
4066: Quad Analog Switch/Multiplexer
4051: 8-Channel Analog Multiplexer/Demultiplexer
4052: Dual 4-Channel Analog Multiplexer/Demultiplexer
4512: 8-Channel Data Selector

4514:	4-Bit Latched 4-to-16 Line Decoder	4040:	12-Bit Binary
4515:	4-Bit Latched 4-to-16 Line Decoder	40160:	Decade Counter with Asynchronous Clear
4529:	Dual 4-Channel or Single 8-Channel Analog Data Selector	40161:	Binary Counter with Asynchronous Clear

Schmitt Triggers
- 4093: Quad 2-Input NAND
- 40106: Hex

Flip-Flops/Latches/Shift Registers
- 4006: 18-Stage Static Shift Register
- 4013: Dual D-Flip-Flop
- 4014: 8-Bit Static Shift Register
- 4015: Dual 4-Bit Static Shift Register
- 4021: 8-Bit Static Shift Register
- 4027: Dual J-K Master/Slave Flip-Flop
- 4031: 64-Stage Static Shift Register
- 4035: 4-Bit Parallel/Parallel Out Shift Register
- 4042: Quad Clocked Latch
- 4043: Quad NOR R/S Latches – Tri-State
- 4044: Quad NAND R/S Latches – Tri-State
- 4076: Quad D-Flip-Flop – Tri-State
- 4094: 8-Stage Shift/Store Register (tri state)
- 40174: Hex D Flip-Flop
- 40175: Quad D Flip-Flop
- 4723: Dual 4-Bit Addressable Latch
- 4724: 8-Bit Addressable Latch

Counters
- 4017: Decade
- 4018: Pre-settable Divide-by-N
- 4020: 14-Bit Binary
- 4022: Octal
- 4024: 7-Stage Ripple
- 4029: Presettable Binary/BCD Up/Down
- 4040: 12-Bit Binary
- 40160: Decade Counter with Asynchronous Clear
- 40161: Binary Counter with Asynchronous Clear
- 40162: Decade Counter with Synchronous Clear
- 40163: Binary Counter with Synchronous Clear
- 40192: 4-Bit Up/Down Decade Counter
- 40193: 4-Bit Up/Down Binary Counter
- 4510: BCD Up/Down Counter
- 4511: BCD-to-7 Segment Latch/Decoder/Driver
- 4516: Binary Up/Down Counter
- 4518: Dual BCD Up
- 4520: Dual Binary Up
- 4522: Programmable Divide-by-N 4-Bit Binary Counter
- 4526: Programmable Divide-by-N 4-Bit Decimal Counter

Oscillators/Timers
- 4541: Programmable Oscillator/Timer

Multivibrators
- 4007: Monostable/Astable Multivibrator
- 4528: Dual Monostable
- 4538: Dual Precision Monostable

Adders/Comparators
- 4008: 4-Bit Full Adder

Other
- 4034: 8-Stage Tri-State Bus Register
- 4041: Quad True/Complement Buffer
- 4046: Phase-Locked Loop
- 4089: Binary Rate Multiplier
- 4543: BCD-to-7 Segment Latch/Decoder/Driver for LCD

4000

Dual 3-Input NOR Gate Plus Inverter

Description: This device is formed by two independent NOR gates and an inverter.

Functional Diagram and/or Package:

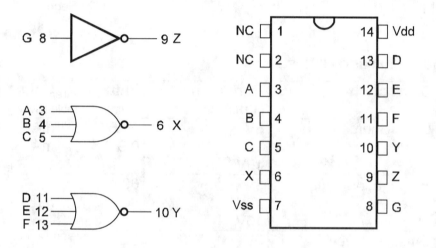

(Figure 2.1)

Pin Names

Vdd – Positive supply voltage (3V to 15V)
Vss – Ground
A, B, C, D, F, G, H – Inputs
X, Y, Z – Outputs

Truth Table:

a) NOR Gates

A	B	C	Out
0	0	0	1
0	0	1	0
0	1	0	0
0	1	1	0
1	0	0	0
1	0	1	0
1	1	0	0
1	1	1	0

b) Inverter

A	Out
0	1
1	0

Operation Mode: The logic signals are applied to the inputs of the gates and the resulted logic levels are taken from the outputs. See the truth table for more details.

Electrical Characteristics:

Characteristic	Conditions (Vdd)	Value (typ)	Units
Drain/Source Current	5V	0.28/0.35	mA
	10V	0.9/0.5	mA
	15V	-	-
Propagation Time Delay	5V	60	ns
	10V	40	ns
	15V	-	-
Quiescent Device Current (max)	5V	0.25	µA
	10V	0.5	µA
	15V	1.0	µA
Supply Voltage Range	3V to 15V		V

Other Devices:

3-Input NOR Gates: 4023, 4025 (buffered). Inverters: 4009 (buffer), 4049 (buffer), 4069

Applications: NOR gates and the inverter can be used in their basic functions adding logic to a project, such as in oscillators, driving stages, and digital amplifiers.

Observations: This device is not recommended for new projects since its functions can be found in other devices of the same family. It is not always easy to find.

Functional Diagrams and Information for Designers

4001

Quadruple 2-Input NOR Gate

Description: This device is formed by four 2-input independent NOR gates in the same package.

Functional Diagram or/and Package:

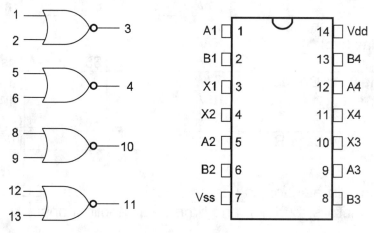

(Figure 2.2)

Pin Names:
  Vdd – Positive Supply Voltage (3V to 15V)
  Vss – Ground
  A1, B1, A2, B2, A3, B3, A4, B4 – Inputs
  X1, X2, X3 – Outputs

Truth Table:

NOR Gate:

A	B	X
0	0	1
0	1	0
1	0	0
1	1	0

Operation Mode:
The logic signals are applied to the inputs and the resulting logic level appears in the output of each gate.

Electrical Characteristics:

Characteristic	Conditions (Vdd)	Value	Units
Drain/Source Current(typ)	5V	0.88	mA
	10V	2.25	mA
	15V	8.8	mA
Propagation Time Delay (typ)	5V	150	ns
	10V	60	ns
	15V	35	ns
Quiescent Device Current (max)	5V	0.25	µA
	10V	0.5	µA
	15V	1.0	µA
Supply Voltage Range	3V to 15V		V

Applications: In addition to use as a logic block, the 4001 can also be used in oscillators, buffers, and time delay applications.

Observations:
- OR gates and inverters can be associated to perform as a NOR gate.
- If the inputs are tied together the logic gates will operate as inverters.

4002

Dual 4-Input NOR Gate

Description: This package contains two independent 4-input NOR gates.

Functional Diagram and/or Package:

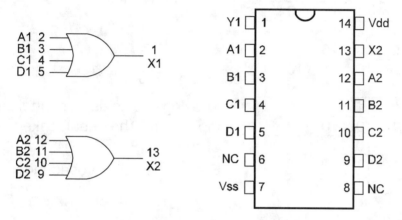

(Figure 2.3)

Pin Names:

Vdd – Positive Supply Voltage (3V to 15V)
Vss – Ground
A1, B1, C1, D1, A2, B2, C2, D2 – Inputs
X1, X2, – Outputs

Truth Table:

A	B	C	D	Out
0	0	0	0	1
0	0	0	1	0
0	0	1	0	0
0	0	1	1	0
0	1	0	0	0
0	1	0	1	0
0	1	1	0	0
0	1	1	1	0
1	0	0	0	0
1	0	0	1	0
1	0	1	0	0
1	0	1	1	0
1	1	0	0	0
1	1	0	1	0
1	1	1	0	0
1	1	1	1	0

Operation Mode:

- The logic signals are applied to the inputs and the result taken from the output.
- Each gate can be used independently.

Electrical Characteristics:

Characteristic	Conditions (Vdd)	Value	Units
Drain/Source Current (typ)	5V	0.88	mA
	10V	2.25	mA
	15V	8.8	mA
Propagation Time Delay (typ)	5V	125	ns
	10V	60	ns
	15V	45	ns
Quiescent Device Current (max)	5V	1	µA
	10V	2	µA
	15V	4	µA
Supply Voltage Range	3V to 15V		V

Applications:

- Logic functions
- Oscillators
- Buffers and drivers

Observations:

The input of each gate can be tied together to form an inverter.

Functional Diagrams and Information for Designers

4006

18-Stage Static Shift Register

Description: This device contains four separated shift register sections – two sections of four stages and two sections of five stages. Each section has an independent data input. The outputs are found in the fourth and fifth stage of each section. The device uses a common clock input for all stages.

Functional Diagram or/and Package:

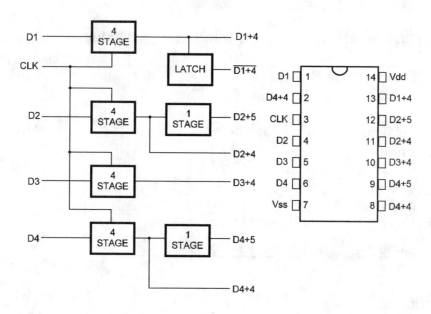

(Figure 2.4)

Pin Names:

Vdd – Positive Supply Voltage (3V to 15V)
Vss – Ground
D1+4, D1+4/, D2+5, D3+4, D4+5 – Outputs
D1, D2, D3, D4 – Inputs CLK – Clock

Truth Table:

D	CLA	D+1
0	↓	0
1	↓	1
X	↑	NC

X – Don't care
↑↓ – Level changes
NC – No change

Operation Mode: A common clock signal is used for all stages. Data is shifted to the next stage on the negative transition of the clock. Through programmed connections of inputs and outputs, multiple register sections of 4, 5, 8, and 9 stages or single sections of 10, 12, 13, 14, 16, 17, and 18 stages can be implemented using one of these devices.

Electrical Characteristics:

Characteristic	Conditions (Vdd)	Value	Units
Drain/Source Current (typ)	5V	0.88	mA
	10V	2.25	mA
	15V	8.8	mA
Maximum Clock Frequency (typ)	5V	2.5	MHz
	10V	5	MHz
	15V	7	MHz
Quiescent Device Current (max)	5V	5	µA
	10V	10	µA
	15V	20	µA
Supply Voltage Range	3V to 15V		V

Applications:

- Shift registers (serial)
- Time delay applications
- Frequency division

Observations:

This device has a low clock input capacitance 6 pF.

4007

Complementary Pair Plus Inverter

Description: This device is formed by three complementary pairs of N- and P-channel MOS transistors. The devices are suitable for series or shunt applications. The input of each pair is protected from static discharge by diode clamps to Vdd and Vss.

Functional Diagram or/and Package:

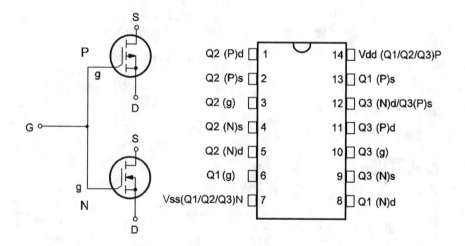

(Figure 2.5)

Pin Names:

Vdd – Positive Supply Voltage
Vss – Ground
D – Drain
G – Gate
S – Source

Truth Table: none

CMOS Sourcebook

Operation Mode: It is recommended for proper operation that the voltages at all pins be constrained to be between Vss −0.3V and Vdd +0.3V. More complex functions can be implemented using multiple packages. Figure 2.6 shows the various possible configurations.

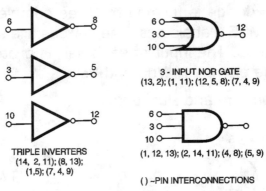

TRIPLE INVERTERS
(14, 2, 11); (8, 13);
(1,5); (7, 4, 9)

3 - INPUT NOR GATE
(13, 2); (1, 11); (12, 5, 8); (7, 4, 9)

(1, 12, 13); (2, 14, 11); (4, 8); (5, 9)

( ) –PIN INTERCONNECTIONS

[Figure 2.6]

Electrical Characteristics:

Characteristic	Conditions (Vdd)	Value	Units
Drain/Source Current (typ)	5V	1.0/4.0	mA
	10V	2.5/2.5	mA
	15V	-	-
Propagation Delay Time (typ)	5V	35	ns
	10V	20	ns
	15V	-	-
Quiescent Device Current (max)	5V	0.05	µA
	10V	0.1	µA
	15V	-	µA
Supply Voltage Range	3V to 15V		V

Applications:

- Digital amplifiers (high impedance input)
- Wave shapers
- Inverters
- Oscillators

4008

4-Bit Full Adder

Description: This package contains four full-adder stages with fast look-ahead, which carries from stage to stage. The parallel carryout bit allows high-speed operation in arithmetic sections using several 4008 inputs.

Functional Diagram and/or Package:

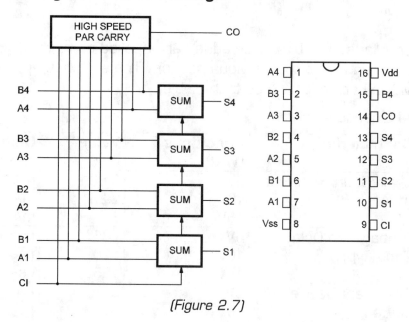

(Figure 2.7)

Pin Names:

Vdd – Positive Supply Voltage (3V to 15V)
Vss – Ground
A1, A2, B1, B2, C1, C2 – Data Input (numbers to be added)
S1, S2, S3, S4 – Sum Bits
CO – Carry Out
CI – Carry In

Truth Table:

A	B	C	CO	SUM
0	0	0	0	0
1	0	0	0	1
0	1	0	0	1
1	1	0	1	0
0	0	1	0	1
1	0	1	1	0
1	1	1	1	1

Operation Mode: The bits to be added are applied to inputs A, B, and C and the carry-in (CI) of the previous section. The sum bits appear in the output with a carryout bit (see truth table).

Electrical Characteristics:

Characteristic	Conditions (Vdd)	Value	Units
Drain/Source Current (typ)	5V	0.88	mA
	10V	2.25	mA
	15V	8.8	mA
Propagation Delay Time (typ)	5V	425	ns
	10V	170	ns
	15V	125	ns
Quiescent Device Current (max)	5V	5	µA
	10V	10	µA
	15V	20	µA
Supply Voltage Range	3V to 15V		V

Applications:

- Arithmetic units

Observations:

The high-speed carryout permits the high-speed operation in arithmetic sections.

4009

Hex Buffers (Inverting)

Description: The six inverting buffers (NOR gates) of this package are independent and can sink 8 mA of current with a Vdd = 10V (typically).

Functional Diagram and/or Package:

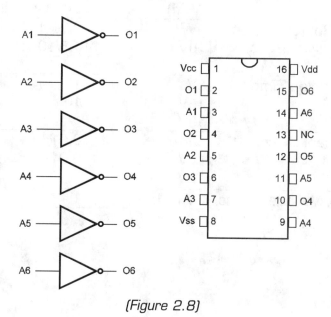

[Figure 2.8]

Pin Names:

Vdd – Positive Supply Voltage (3V to 15V)
Vss – Ground
Vcc – Output Voltage (Vcc<Vdd)
A1, A2, A3, A4, A5, A6 – Inputs
O1, O2, O3, O4, O5, O6 – Outputs
NC – Not Connected

Truth Table:

A	0
0	1
1	0

Operation Mode: All the buffers are independent. The output depends on the logic applied to the input according the truth table.

Electrical Characteristics:

Characteristic	Conditions (Vdd)	Value	Units
Drain/Source Current (typ)	5V	2.25/4	mA
	10V	4.5/10	mA
	15V	-	-
Propagation Time Delay	5V	15	ns
	10V	10	ns
	15V	-	-
Quiescent Device Current (max)	5V	0.01	µA
	10V	0.01	µA
	15V	-	-
Supply Voltage Range	3V to 15V		V

Applications:

- Digital amplifiers
- CMOS to TTL interfaces
- High-to-Low Logic Converter
- Multiplexer (1 to 6 or 6 to 1)

Observations:

These buffers may be used as hex buffers in CMOS for TTL interfacing.

4010

Hex Buffers (Non-Inverting)

Description: The six non-inverting buffers found in this package can be used independently and have high output current capabilities. They can drain 8 mA with a Vdd of 10V (typically).

Functional Diagram and/or Package:

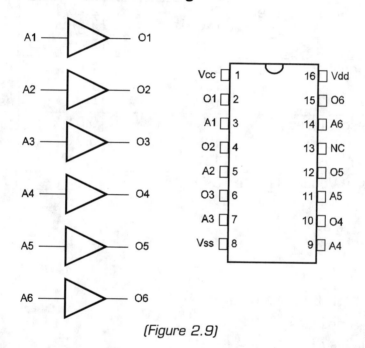

(Figure 2.9)

Pin Names:

Vdd – Positive Supply Voltage (3V to 15V)
Vss – Ground
Vcc – Output Voltage (Vcc<Vdd)
A1, A2, A3, A4, A5, A6 – Inputs
O1, O2, O3, O4, O5, O6 – Outputs
NC – Not Connected

Truth Table:

A	0
0	1
1	0

Operation Mode: All buffers in this package are independent. The output logic level depends on the logic levels applied to the inputs according the truth table.

Electrical Characteristics:

Characteristic	Conditions (Vdd)	Value (typ)	Units
Drain/Source Current	5V	2.25/4.0	mA
	10V	4.5/10	mA
	15V	-	-
Propagation Time Delay	5V	15	ns
	10V	10	ns
	15V	-	-
Quiescent Device Current (max)	5V	0.01	µA
	10V	0.01	µA
	15V	-	-
Supply Voltage Range	3V to 15V		V

Applications:

- CMOS to TTL interfacing
- Digital amplifiers
- Multiplexer (1 to 6 or 6 to 1)

Observations: The buffers in this package may be used as hex buffers CMOS-to-TTL or as current drivers.

Functional Diagrams and Information for Designers

4011

Quadruple 2-Input NAND Gate

Description: This device is formed by four independent 2-input NAND gates.

Functional Diagram and/or Package:

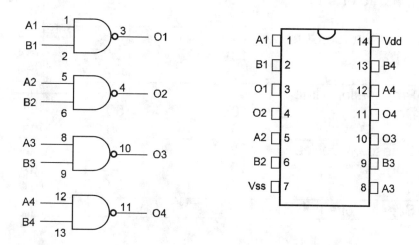

(Figure 2.10)

Pin Names:

Vdd – Positive Supply Voltage
Vss – Ground
A1, B1, A2, B2, A3, B3, A4, B4 – Inputs
O1, O2, O3, O4 – Outputs

Truth Table:

A	B	O
0	0	1
0	1	1
1	0	1
1	1	0

Operation Mode: All the gates inside this package are independent. The logic level at the output of each gate depends on the logic signals applied to the inputs according the truth table.

Electrical Characteristics:

Characteristic	Conditions (Vdd)	Value	Units
Drain/Source Current (typ)	5V	0.88	mA
	10V	2.25	mA
	15V	8.8	mA
Propagation Time Delay (typ)	5V	120	ns
	10V	50	ns
	15V	35	ns
Quiescent Device Current (max)	5V	0.004	µA
	10V	0.005	µA
	15V	0.006	µA
Supply Voltage Range	3V to 15V		V

Applications:

- Logic functions
- Oscillators
- Digital amplifier (driver)

Observations:

The inputs can be wired together to form inverters. An inverter can also be obtained connecting one input to the Vdd.

4012

Dual 4-Input NAND Gate

Description:

Two independent 4-input NAND gates can be found in this package.

Functional Diagram and/or Package:

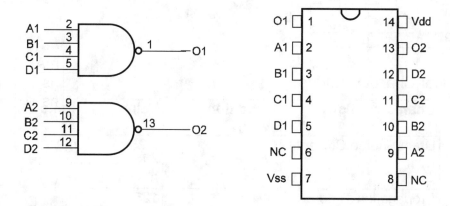

(Figure 2.11)

Pin Names:

Vdd – Positive Supply Voltage (3V to 15V)
Vss – Ground
A1, B1, C1, D1, A2, B2, C2, D2 – Inputs
O1, O2 – Outputs
NC – Not Connected

Truth Table:

A	B	C	D	O
0	0	0	0	0
0	0	0	1	0
0	0	1	0	0
0	0	1	1	0
0	1	0	0	0
0	1	0	1	0
0	1	1	0	0
0	1	1	1	0
1	0	0	0	0
1	0	0	1	0
1	0	1	0	0
1	0	1	1	0
1	1	0	0	0
1	1	0	1	0
1	1	1	0	0
1	1	1	1	1

Operation Mode: All the gates inside this package are independent. The logic level at the output of each gate depends on the logic levels applied to the inputs according the truth table.

Electrical Characteristics:

Characteristic	Conditions (Vdd)	Value	Units
Drain/Source Current (typ)	5V	2.25	mA
	10V	4.5	mA
	15V	8.8	mA
Propagation Time Delay (typ)	5V	125	ns
	10V	60	ns
	15V	45	ns
Quiescent Device Current (max)	5V	0.25	µA
	10V	0.5	µA
	15V	1.0	µA
Supply Voltage Range	3V to 15V		V

Applications:

- Logic functions
- Digital amplifiers (driver)
- Oscillators

Observations:

The inputs can be connected together to form inverters.

4013

Dual D Flip-Flop

Description: This device is formed by two independent D-type flip-flops. Each flip-flop has its own data, set, reset, and clock inputs. Each flip-flop has normal and complementary outputs.

Functional Diagram and/or Package:

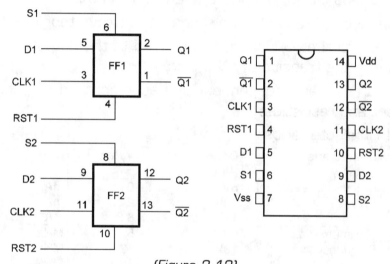

(Figure 2.12)

Pin Names:

    Vdd – Positive Supply Voltage
    Vss – Ground
    Q1, Q2 – Outputs
    Q1/, Q2/ – Complementary Outputs
    CLK1, CLK2 – Clocks
    RST1, RST2 – Reset
    S1, S2 – Set
    D1, D2 – Data

Truth Table:

CL	D	R	S	Q	Q/
↑	0	0	0	0	1
↑	1	0	0	1	0
↓	X	0	0	Q	Q/
X	X	1	0	0	1
X	X	0	1	1	0
X	X	1	1	1	1

X – no change
↓↑ – Level changes

Operation Mode:

- The flip-flops inside this package are independent.
- Connecting Q/ to D of each stage – the circuit toggle.
- The logic level at D input is transferred to the output Q during the positive going transition of the clock.
- Set and reset are independent and accomplished by a logic "1" on the set and reset lines.

Electrical Characteristics:

Characteristic	Conditions (Vdd)	Value	Units
Drain/Source Current (typ)	5V	0.88	mA
	10V	2.25	mA
	15V	8.8	mA
Maximum Clock Frequency (typ)	5V	5	MHz
	10V	12.5	MHz
	15V	15.5	MHz
Quiescent Device Current (max)	5V	1.0	µA
	10V	2.0	µA
	15V	4.0	µA
Supply Voltage Range	3V to 15V		V

Applications:

- Control circuits
- Registers
- Counters

Observations:

Setting and resetting are independent of the clock and accomplished by a high level on the set or reset line respectively.

4014

8-Stage Static Shift Register

Description: This package contains an 8-stage parallel input/serial output shift register. Each of the 8 stages has its own JAM input. The outputs are available from the sixth, seventh, and eighth stages. All the outputs can sink or drain the same current.

Functional Diagram and/or Package:

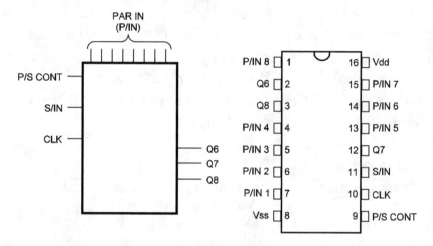

(Figure 2.13)

Pin Names:

Vdd – Positive Supply Voltage (3V to 15V)

Vss – Ground

P/IN (1 to 8) – Parallel Input

S/IN – Serial IN

CLK – Clock

P/S CONT – Parallel/Serial Control

BUF OUT (Q6, Q7, Q8) – Buffered Outputs

Truth Table:

CL	SI	P/S CTRL	PI1	Pin	Q1 (internal	Qn
↑	X	1	0	0	0	0
↑	X	1	1	0	1	0
↑	X	1	0	1	0	1
↑	X	1	1	1	1	1
↑	0	0	X	0	0	Qn-1
↑	1	0	X	X	1	Qn-1
↓	X	X	X	X	Q1	Qn

X = Don't care

Operation Mode:

- A parallel/serial control enables individual JAM inputs to each of the 8 stages. The outputs Q are available from the 6th and 7th stages.
- Data is serially shifted into the register synchronously with the positive transition of the clock when the P/S input is "0".
- When the P/S input is "1" data is jammed into each stage synchronously with the positive transition of the clock. (See the truth table for more details.)

Electrical Characteristics:

Characteristic	Conditions (Vdd)	Value (typ)	Units
Drain/Source Current	5V	0.88	mA
	10V	2.2	mA
	15V	8.0	mA
Maximum Clock Frequency (typ)	5V	4	MHz
	10V	12	MHz
	15V	16	MHz
Quiescent Device Current (typ) (max)	5V	0.1	µA
	10V	0.2	µA
	15V	0.3	µA
Supply Voltage Range	3V to 15V		V

Applications:

- Parallel to Serial Data Conversion
- General Purpose Register
- Data Queuing (PISO – Parallel In/Serial Out)

4015

Dual 4-Bit Static Shift Register

Description: Two identical 4-Bit Static Shift Registers can be found in this package. They have serial input/parallel output registers with independent Data, Clock, and Reset.

Functional Diagram and/or Package:

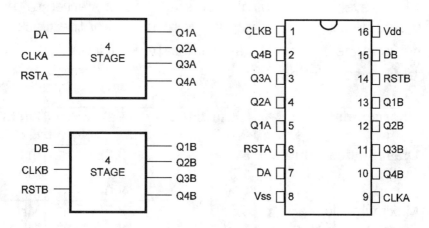

(Figure 2.14)

Pin Names:

Vdd – Positive Supply Voltage (3V to 15V)
Vss – Ground
DA, DB – Data
CLKA, CLKB – Clock
RSTA, RSTB – Reset
Q1A, Q2A, Q3A, Q4A, Q1B, Q2B, Q3B, Q4B – Outputs

Truth Table:

CL	D	R	Q1	Qn
↑	0	0	0	Qn-1
↑	1	0	1	Qn-1
↓	X	0	Q1	Qn
X	X	1	0	0

X – Don't care
↑↓ – Level changes

Operation Mode:

- Each of the two shift registers found in this package can be used independently since they have separated Data, Clock and Reset inputs.
- The logic level in the input of each stage is transferred to the output of the same stage with the positive transition of the clock.
- Reset is made with an "1" applied at the RST input.

Electrical Characteristics:

Characteristic	Conditions (Vdd)	Value	Units
Drain/Source Current (typ)	5V	0.88	mA
	10V	2.25	mA
	15V	8.8	mA
Maximum Clock Frequency (typ)	5V	3.5	MHz
	10V	8	MHz
	15V	11	MHz
Quiescent Device Current (max)	5V	5	µA
	10V	10	µA
	15V	15	µA
Supply Voltage Range	3V to 15V		V

Applications:

- General Purpose Shift Register
- Data Conversion (Serial to Parallel)
- Data Queuing (SIPO O Serial In/Parallel Out)

Observations:

The inputs are protected from static discharge.

Functional Diagrams and Information for Designers

4016

Quad Bilateral Switch

Description: This package contains four independent bilateral switches intended for the transmission, control, or multiplexing of analog/digital signals.

Functional Diagram and/or Package:

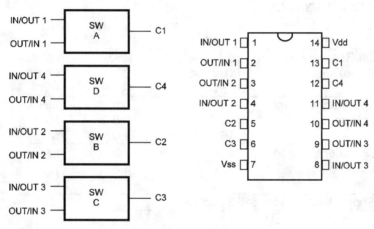

(Figure 2.15)

Pin Names:

   Vdd – Positive Supply Voltage (3V to 15V)
   Vss – Ground
   IN/OUT1, IN/OUT2, IN/OUT3, IN/OUT4 – Inputs/Outputs
   OUT/IN1, OUT/IN2, OUT/IN3, OUT/IN4 – Output/Inputs
   C1, C2, C3, C4 – Control Inputs

Truth Table: none

Operation Mode:

- All switches found in this package are independent.
- When operating with digital signals, Vdd is sourced to pin 14 and pin 7 is wired to the ground (GND).
- When operating with analog signals, pin 14 is supplied with +5V and pin 7 with -5V. The amplitude of the analog signals can't be higher than 10Vpp (-5 to +5V).
- The switches are on when the control input is at the logic level "1" and off when the logic level is "0".

Electrical Characteristics:

Characteristic	Conditions (Vdd)	Value	Units
On Resistance (Ron) – (typ)	5V	250	Ohms
	10V	200	Ohms
	15V	-	-
Maximum Control Frequency (typ)	5V	6.5	MHz
	10V	8.0	MHz
	15V	9.0	MHz
Quiescent Device Current (max)	5V	0.25	µA
	10V	0.5	µA
	15V	1.0	µA
Supply Voltage Range	3V to 15V		V

Other Devices:

The 4066 also have 4 analog/digital switches. (See Observations section.)

Applications:

- Modulator/demodulator
- Squelch Circuits
- Multiplexing
- Commutation
- Chopper
- Signal Gating
- Logic implementation
- Digital control (frequency, impedance, phase, gain, etc.)

Observations:

- This device is pin-for-pin compatible with the 4066, which has a lower ON resistance and is recommended for new projects.
- The OFF resistance is very high, typically 10^{12} ohms.

4017

Decade Counter/Divider with 10 Decoded Outputs

Description: This device is a 5-stage divide-by-10 Johnson Counter with 10 decoded outputs and a carry out bit. The decoded output goes to the "1" logic level remaining the others in the "0" logic level.

Functional Diagram and/or Package:

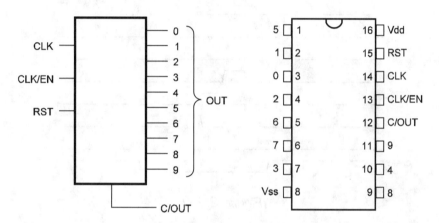

(Figure 2.16)

Pin Names:

Vdd – Positive Supply Voltage (3V to 15V)
Vss – Ground
OUT0 to OUT9 – Decoded Outputs
RST – Reset
CLK – Clock or Signal Input
CLK EN – Clock Enable
C/OUT – Carry Out Output

CMOS Sourcebook

Timing diagrams: Instead of the truth table, the operation of the 4017 can be clarified by looking at the timing diagram shown in Figure 2.17.

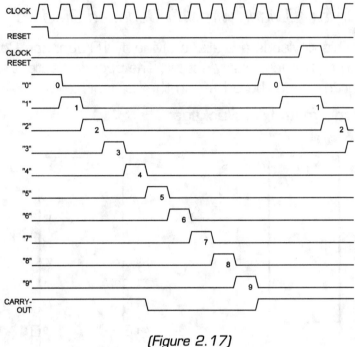

(Figure 2.17)

Operation Mode:

- In the normal operation RST and CLK EN are kept at the ground (0) and the pulses are applied to the CLK input.
- The circuit advances one count on the positive going transition of the clock pulse or signal.
- Reset is done by putting a "1" at the RST input.
- Each decoded output remains at the "1" logic level during the full clock cycle. The other remains at the "0" logic level.
- The carryout pulse is produced at the final of the full counting.
- The 4017 can be programmed to count up to values lower than 10 (2 and 9). To count up to N, simply connect the output n+1 to the RST as shown in Figure 2.18.

Functional Diagrams and Information for Designers

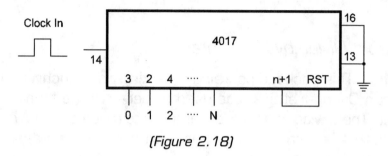

(Figure 2.18)

Electrical Characteristics:

Characteristic	Conditions (Vdd)	Value	Units
Drain/Source Current (typ)	5V	0.88	mA
	10V	2.25	mA
	15V	8.8	mA
Maximum Clock Frequency (typ)	5V	2	MHz
	10V	5	MHz
	15V	6	MHz
Quiescent Device Current (max)	5V	0.3	µA
	10V	0.5	µA
	15V	1.0	µA
Supply Voltage Range	3V to 15V		V

Applications:

- Decade Counters (sequencer)
- Binary Counter with Decoder
- Frequency Dividers
- Timers
- Divide-by-N Counters

4018

Presettable Divide-by-N Counter

Description: This package contains a walking-ring synchronous counter. Six Johnson Counter stages are used to perform the functions found in this circuit. The device can be programmed to divide an input by 2 through 10. The output is a square wave for even divisions and nearly a square wave for odd divisions.

Functional Diagram and/or Package:

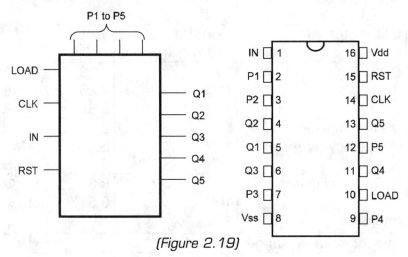

(Figure 2.19)

Pin Names:

Vdd – Positive Supply Voltage (3V to 15V)

Vss – Ground

P1 to P5 – Parallel Load Inputs

Q1 to Q5 – Outputs

CLK or DATA – Clock (signal input)

RST – Reset

LOAD – Load Input also called Preset Enable

IN – Feedback Input

Functional Diagrams and Information for Designers

Timing Diagram:

The operation of this device is better understood if the timing diagram is given replacing the truth table. The timing diagram is shown in Figure 2.20.

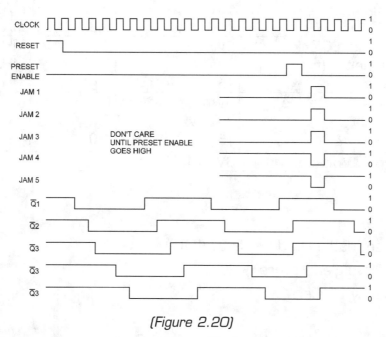

(Figure 2.20)

Operation Mode

- The program is made by changing the feedback according to the table below:

Interconnecting Pins	Divide by:
Q1 to IN	2
Q1 and Q2 to IN	3
Q2 and IN	4
Q2 and Q3 to IN	5
Q3 to IN	6
Q3 and Q4 to IN	7
Q4 to IN	8
Q4 and Q5 to IN	9
Q5 to IN	10

- In normal operation Reset and Load are put to the ground. The counter advances one count for each positive transition of the clock. The outputs Q0 to Q5 are buffered.
- Reset – To reset the circuit, place the Reset input to the "1" logic level.
- Load – Parallel load is created by making the Load input positive.

Electrical Characteristics:

Characteristic	Conditions (Vdd)	Value	Units
Drain/Source Current (typ)	5V	0.88	mA
	10V	2.25	mA
	15V	8.8	mA
Maximum Clock Frequency (typ)	5V	4	MHz
	10V	9	MHz
	15V	14	MHz
Quiescent Device Current (max)	5V	0.3	µA
	10V	0.5	µA
	15V	1.0	µA
Supply Voltage Range	3V to 15V		V

Applications:

- Programmable Dividers (2 to 10)
- Sine Wave Generators
- Frequency Dividers
- Timers
- Programmable Decade Counters and Greater

Observations:

- The parallel code is specialized.

4019

Quad AND-OR Select Gate

Description: This device acts as a 4-pole double throw switch. Inside the package four AND-OR select gates with common select logic are found.

Functional Diagram and/or Package:

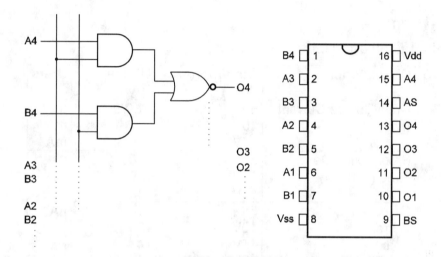

(Figure 2.21)

Pin Names:

Vdd – Positive Supply Voltage (3V to 15V)
Vss – Ground
AS, BS – Selection Inputs
A1, B1, A2, B2, A3, B3, A4, B4 – Inputs
O1, O2, O3, O4 – Outputs

Truth Table:

SA	SB	A	B	O
1	0	1	X	1
1	0	0	X	0
0	1	X	1	1
0	1	X	0	0
0	0	X	X	0
1	1	0	0	0
1	1	0	1	1
1	1	1	0	1
1	1	1	1	1

Operation Mode:

- The function selection is made by the inputs AS and SB according the truth table.

X – Don't care

Electrical Characteristics:

Characteristic	Conditions (Vdd)	Value (typ)	Units
Drain/Source Current (typ)	5V	1.0/0.4	mA
	10V	2,5/1.0	mA
	15V	10/3.0	mA
Propagation Delay Time (typ)	5V	100	ns
	10V	50	ns
	15V	45	ns
Quiescent Device Current (max)	5V	1	µA
	10V	2	µA
	15V	4	µA
Supply Voltage Range	3V to 15V		V

Other Devices: The 4519 is similar, but has the option of an exclusive OR function.

Applications:

- Select Gating (AND-OR)
- AND-OR/Exclusive OR Selectors
- True/Complement Selectors
- Shift Left/Shift Right Registers

Observations:

This circuit acts only as a data selector – it isn't a data distributor.

Functional Diagrams and Information for Designers

4020

14-Stage Ripple Carry Binary Counter

Description: This device contains a 14-stage binary counter that counts upward. The circuit can be used as a frequency divider up to 16,384. The circuit has no outputs for the second and third stages.

Functional Diagram and/or Package:

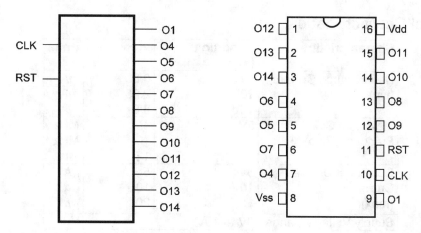

(Figure 2.22)

Pin Names:

Vdd – Positive Supply Voltage (3V to 15V)

Vss – Ground

O1, O3, O4, O4, O6, O7, O8, O9, O10, O11, O12, O13, O14 – Outputs

RST – Reset

CLK – Clock or Input

Truth Table: none

Operation Mode:

- RST is connected to the ground in normal applications.
- The circuit advances one count at each negative input pulse. The division is made by a number corresponding to a power of 2 of the considered output. For instance, in the output O4 we have a division by $2^4 = 16$. The maximum division is found in the 14th output (O14) where the division quotient is $2^{14} = 16,384$.
- To force all the outputs to the "O" level, make the RESET input positive.

Electrical Characteristics:

Characteristic	Conditions (Vdd)	Value	Units
Drain/Source Current (typ)	5V	0.88	mA
	10V	2.25	mA
	15V	8.8	mA
Maximum Clock Frequency (typ)	5V	4	MHz
	10V	10	MHz
	15V	12	MHz
Quiescent Device Current (max)	5V	5	µA
	10V	10	µA
	15V	20	µA
Supply Voltage Range	3V to 15V		V

Other Devices:

- The 4040 is a divider with 12 stages dividing for values up to 4,096.
- The 4060 is a divider with 14 stages dividing for values up to 16,384. The 4060 has an internal oscillator.

Applications:

- Timers
- Frequency Dividers
- Counters
- Time Delay Applications

Observations:

This device is a ripple counter. When setting the time, incorrect counting can result.

4021

8-Stage Static Shift Register

Description: This package contains a Parallel-In/Serial-Out Shift Register that can be used as a 6, 7, or 8 stage shift-right register, either as a serial in/serial out or as a parallel in/serial out circuit.

Functional Diagram and/or Package:

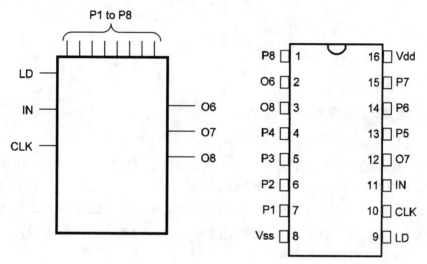

(Figure 2.23)

Pin Names

Vdd – Positive Supply Voltage
Vss – Ground
O6, O7, O8 – Outputs
P1, P2, P3, P4, P5, P6, P7, P8 – Parallel Inputs
CLK – Clock
IN – Data Input
LD – Load

Truth Table:

CL	SI	P/S Ctrl	PI1	PIn	Q1 (internal)	Qn
X	X	1	0	0	0	0
X	X	1	0	1	0	1
X	X	1	1	0	1	0
X	X	1	1	1	1	1
↑	0	0	X	X	0	Qn-1
↑	1	0	X	X	1	Qn-1
↓	X	1	X	X	Q1	Qn

↓↑ - Level changes
X – Don't care

Operation Mode:

a) Serial-In/Serial-Out Operation:

- The LD input is grounded.
- Data applied in the IN input is shifted to the first stage on the positive transition of the clock pulse.
- After six successive clock cycles, data appears at the output pin Q6.
- In the next clock pulse, the data goes to the output Q7, and yet another, to the output Q8.
- Additional clock pulses lose the data or recirculate it if stages are cascaded.

b) Parallel-In/Serial-Out Operation:

- An 8-bit word is loaded through the parallel inputs P1 to P8 (P1 near the input and P8 near the output).
- When LD goes to the "1" logic level, data is loaded into the register. After this operation, LD must be returned to ground.
- The pulses applied to the clock shift the data loaded into the register to the right.

Functional Diagrams and Information for Designers

Electrical Characteristics:

Characteristic	Conditions (Vdd)	Value	Units
Drain/Source Current (typ)	5V	0.88	mA
	10V	2.21	mA
	15V	8.0	mA
Maximum Clock Frequency (typ)	5V	3.5	MHz
	10V	10	MHz
	15V	16	MHz
Quiescent Device Current (max)	5V	0.1	µA
	10V	0.2	µA
	15V	0.3	µA
Supply Voltage Range	3V to 15V		V

Other Devices:

- The 4014 is a similar device with synchronous load.

Applications:

- Registers
- Data Conversion

Observations:

Fast clock transitions are needed to avoid incorrect operation of the device.

4022

Divide-by-8 Counter/Divider

Description: This package contains a full Divide-by-8 Counter. This device is synchronous and produces a square wave as output.

Functional Diagram and/or Package:

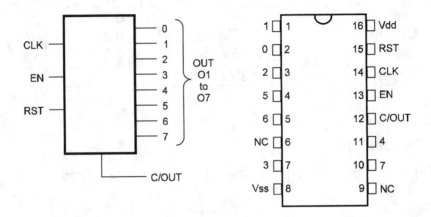

(Figure 2.24)

Pin Names:

Vdd – Positive Supply Voltage
Vss – Ground
O0, O1, O2, O3, O4, O5, O6, O7 – Outputs
CLK – Clock or Input
EN – Enable
RST – Reset
C/OUT – Output (see Operation Mode)

Functional Diagrams and Information for Designers

Timing Diagram: Instead of the truth table, a timing diagram is given for this device.

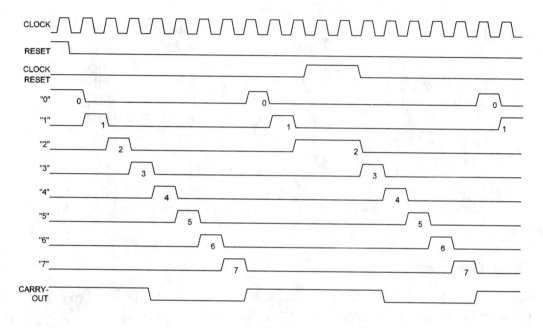

(Figure 2.25)

Operation Mode:

- EN and RST must be grounded.
- Any positive transition of the clock makes the circuit advance one count.
- The decoded output goes to the "1" logic level (positive) and the others remain at the "0" state (ground).
- The OUT output remains in the "1" level for counting from 0 to 3 and goes to the "0" logic level when counting from 4 through 7.
- Reset is completed by putting the RST terminal to the "1" logic level. RST must be back to the "0" to allow the count to continue.
- A "1" applied to the EN input inhibits the clock operation and stops the count.

Electrical Characteristics:

Characteristic	Conditions (Vdd)	Value	Units
Drain/Source Current (typ)	5V	0.88	mA
	10V	2.25	mA
	15V	8.8	mA
Maximum Clock Frequency (typ)	5V	2	MHz
	10V	5	MHz
	15V	6	MHz
Quiescent Device Current (max)	5V	0.3	µA
	10V	0.5	µA
	15V	1.0	µA
Supply Voltage Range	3V to 15V		V

Other Devices:

The 4017 is a similar counter, but it presents 10 outputs. With appropriate connections, it also can be used to count up to 8.

Applications:

- Counters
- Frequency Dividers
- Timers

Observations:

External gates can be used to make divisions from 1 to 8.

4023

Triple 3-Input NAND Gate

Description: Each of the three 3-Input NAND gates found in this package can be used independently.

Functional Diagram and/or Package:

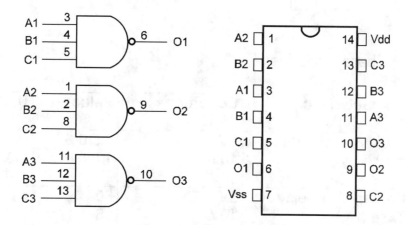

(Figure 2.26)

Pin Names:

Vdd – Positive Supply Voltage (3V to 15V)
Vss – Ground
A1, B1, C1, A2, B2, C2, A3, B3, C3 – Inputs
O1, O2, O3 – Outputs

Truth Table:

A	B	C	O
0	0	0	1
0	0	1	1
0	1	0	1
0	1	1	1
1	0	0	1
1	0	1	1
1	1	0	1
1	1	1	0

Operation Mode:

- Logic levels are applied to the inputs resulting in a logic level at the output as shown by the truth table.

Electrical Characteristics:

Characteristic	Conditions (Vdd)	Value	Units
Drain/Source Current (typ)	5V	0.5/0.5	mA
	10V	1.0/0.6	mA
	15V	-	-
Propagation Delay Time (typ)	5V	50	ns
	10V	25	ns
	15V	-	ns
Quiescent Device Current (max)	5V	0.05	µA
	10V	0.1	µA
	15V	-	µA
Supply Voltage Range	3V to 15V		V

Applications:

- Logic Functions
- Oscillators
- Digital Amplifiers

Observations:

The device can be used as a triple 2-Input NAND gate if one of the inputs of each gate remains at the "1" logic level.

4024

7-Stage Ripple Carry Binary Counter

Description: This package contains a 7-stage binary ripple counter that can also be used as divide-by-128 circuit. The circuit is counted up using positive logic.

Functional Diagram and/or Package:

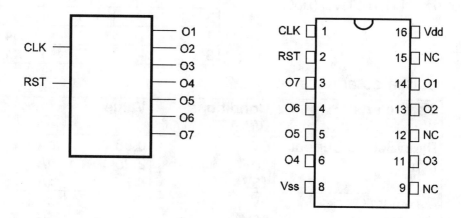

(Figure 2.27)

Pin Names:

Vdd – Positive Supply Voltage (3V to 15V)
Vss – Ground
O1, O2, O3, O4, O5, O6, O7 – Outputs
CLK – Clock or Input
RST – Reset
NC – Not Connected

Truth Table: none

Operation Mode:

- RST is held at ground.
- The circuit advances one count in the positive transition of the clock signal.
- The outputs divide the clock signal by powers of two according the table on the right:
- RST is done by putting the pin to the "1" logic level (positive).

Output Pin Number	n	2^n
12	1	2
11	2	4
9	3	8
6	4	16
5	5	32
4	6	64
3	7	128

Electrical Characteristics:

Characteristic	Conditions (Vdd)	Value	Units
Drain/Source Current (typ)	5V	0.88	mA
	10V	2.25	mA
	15V	8.8	mA
Maximum Clock Frequency (typ)	5V	5	MHz
	10V	12	MHz
	15V	15	MHz
Quiescent Device Current (max)	5V	0.3	µA
	10V	0.5	µA
	15V	0.7	µA
Supply Voltage Range	3V to 15V		V

Applications:

- Counters
- Frequency Dividers
- Time Delay Circuits

Observations:

This is a ripple counter. When setting the time, the circuit counts incorrectly.

Functional Diagrams and Information for Designers

4025

Triple 3-Input NOR Gate

Description: This device is formed by three independent 3-Input NOR gates.

Functional Diagram and/or Package:

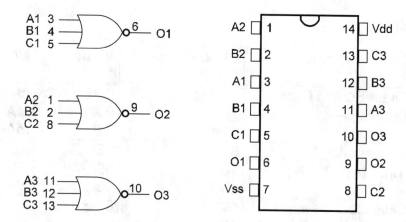

(Figure 2.28)

Pin Names:

Vdd – Positive Supply Voltage (3V to 15V)
Vss – Ground
A1, B1, C1, A2, B2, C2, A3, B3, C3 – Inputs
O1, O2, O3 – Outputs

Truth Table:

A	B	C	O
0	0	0	1
0	0	1	0
0	1	0	0
0	1	1	0
1	0	0	0
1	0	1	0
1	1	0	0
1	1	1	0

Operation Mode:

- The logic signals are applied to the inputs and in the output. The result is given according the truth table.

Electrical Characteristics:

Characteristic	Conditions (Vdd)	Value	Units
Drain/Source Current (typ)	5V	0.88	mA
	10V	2.25	mA
	15V	8.8	mA
Propagation Delay Time (typ)	5V	130	ns
	10V	60	ns
	15V	40	ns
Quiescent Device Current (max)	5V	0.25	µA
	10V	0.5	µA
	15V	1.0	µA
Supply Voltage Range	3V to 15V		V

Other Devices:

The 4001 is a Quad 2-Input NOR Gate.

Applications:

- Logic Functions
- Digital Amplifiers
- Oscillators

Observations:

Any gate can be used as a 2-Input NOR gate if the remaining input is put to ground.

Functional Diagrams and Information for Designers

4027

Dual J-K Master/Slave Flip-Flop with Set and Reset

Description: This package contains two independent J-K flip-flops. Each clock can operate in two modes: direct and clocked.

Functional Diagram and Package:

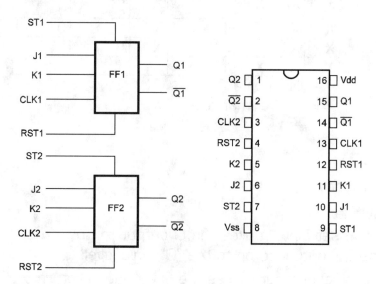

(Figure 2.29)

Pin Names:

Vdd – Positive Supply Voltage (3V to 15V)
Vss – Ground
ST1, ST2 – Set
RST1, RST2 – Reset
CLK1, CLK2 – Clock
J1, J2 – J Inputs
K1, K2 – K Inputs
Q1, Q2 – Outputs Q1/, Q2/ – Complementary Outputs

Truth Table:

CL	J	K	S	R	Q	Q	Q/
↑	1	X	0	0	0	1	0
↑	X	0	0	0	1	1	0
↑	0	X	0	0	0	0	1
↑	X	1	0	0	1	0	1
↓	X	X	0	0	X	No change	No change
X	X	X	1	0	X	1	0
X	X	X	0	1	X	0	1
X	X	X	1	1	X	1	1

X – Don't care
↓↑ - Level changes

Operation Mode:

a) Direct Mode
- A positive set input makes the Q output to go to the "1" logic level and the Q/ output to go to the "0" logic level.
- A positive reset input makes the Q output to return to the "0" logic level and the Q/output to the "1" logic level.
- If positive inputs are applied simultaneously to both inputs (ST and RST), Q and Q/ go to the "1" logic level. This is a disallowed state.

b) Clocked Mode
- Set and Clear inputs remains at ground.
- The logic signals applied to the J and K inputs determine the final state of the outputs of the flip-flop.
- The changes in the states of the flip-flop occur with the positive transition of the clock signal.

Functional Diagrams and Information for Designers

Electrical Characteristics:

Characteristic	Conditions (Vdd)	Value	Units
Drain/Source Current (typ)	5V	0.88	mA
	10V	2.25	mA
	15V	8.8	mA
Maximum Clock Frequency (typ)	5V	6	MHz
	10V	12.5	MHz
	15V	15.5	MHz
Quiescent Device Current (max)	5V	1	µA
	10V	2	µA
	15V	4	µA
Supply Voltage Range	3V to 15V		V

Applications:

- Counters
- Registers
- Control Applications

Observations:

All inputs are protected against damage due to electrostatic discharges.

4028

BCD-to-Decimal Decoder

Description: This device is formed by a circuit that converts Binary-Coded-Decimal into a 1-of-10 output. The same device can be used to convert a 3-bit code into a 1-of-8 output.

Functional Diagram or/and Package:

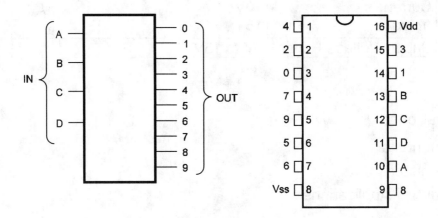

(Figure 2.30)

Pin Names:

Vdd – Positive Supply Voltage (3V to 15V)
Vss – Ground
A, B, C, D – BCD Inputs
0, 1, 2, 3, 4, 5, 6, 7, 8, 9 – Decimal Outputs

Functional Diagrams and Information for Designers

Truth Table:

D	C	B	A	0	1	2	3	4	5	6	7	8	9
0	0	0	0	1	0	0	0	0	0	0	0	0	0
0	0	0	1	0	1	0	0	0	0	0	0	0	0
0	0	1	0	0	0	1	0	0	0	0	0	0	0
0	0	1	1	0	0	0	1	0	0	0	0	0	0
0	1	0	0	0	0	0	0	1	0	0	0	0	0
0	1	0	1	0	0	0	0	0	1	0	0	0	0
0	1	1	0	0	0	0	0	0	0	1	0	0	0
0	1	1	1	0	0	0	0	0	0	0	1	0	0
1	0	0	0	0	0	0	0	0	0	0	0	1	0
1	0	0	1	0	0	0	0	0	0	0	0	0	1
1	0	1	0	0	0	0	0	0	0	0	0	1	0 *
1	0	1	1	0	0	0	0	0	0	0	0	0	1 *
1	1	0	0	0	0	0	0	0	0	0	0	1	0 *
1	1	0	1	0	0	0	0	0	0	0	0	0	1 *
1	1	1	0	0	0	0	0	0	0	0	0	1	0 *
1	1	1	1	0	0	0	0	0	0	0	0	0	1 *

* Extraordinary states

Operation Mode:

a) BCD to 1-of-10
- BCD codes are applied to the A, B, C, and D inputs. The least significant bit is applied to A and the most significant to D.
- The output that goes to the "1" logic level depends on the input (see the truth table).
- An invalid state depends on the origin of the chip (manufacturer).

b) 1-of-8 decoder
- The input D is grounded.
- The logic signals to be decoded are applied to inputs A, B, and C.
- The output depends on the logic levels applied to the input (see the truth table).

Electrical Characteristics:

Characteristic	Conditions (Vdd)	Value	Units
Drain/Source Current (typ)	5V	1.0/0.4	mA
	10V	2.6/1.0	mA
	15V	8.8/3.0	mA
Propagation Delay Time (typ)	5V	240	ns
	10V	100	ns
	15V	70	ns
Quiescent Device Current (max)	5V	1	µA
	10V	2	µA
	15V	4	µA
Supply Voltage Range	3V to 15V		V

Applications:

- Code Conversion
- Address Decoding – Memory Controls
- Driver for Displays (tubes)

Observations:

Outputs can be relabeled for other codes.

4029

Presettable Binary/Decade Up/Down Counter

Description: The circuit inside this package may be programmed to be a divide-by-10 or a divide-by-16 (decade or hexadecimal) counter. The circuit is synchronous and advances one count in the positive going input (clock) pulse.

Functional Diagram and/or Package:

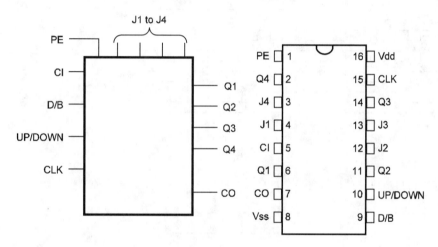

(Figure 2.31)

Pin Names:

Vdd – Positive Supply Voltage (3V to 15V)
Vss – Ground
CLK – Clock
Q1, Q2, Q3, Q4 – Outputs
J1, J2, J3, J4 – JAM Inputs

PE – Preset Enable
CI – Carry In
CO – Carry Out
UP/DOWN – Up-Down Selection
D/B – Decade/Binary Selection

CMOS Sourcebook

Waveforms: Figure 2.32 shows the logic waveforms in the circuit operating in the two possible modes: decade mode and binary mode.

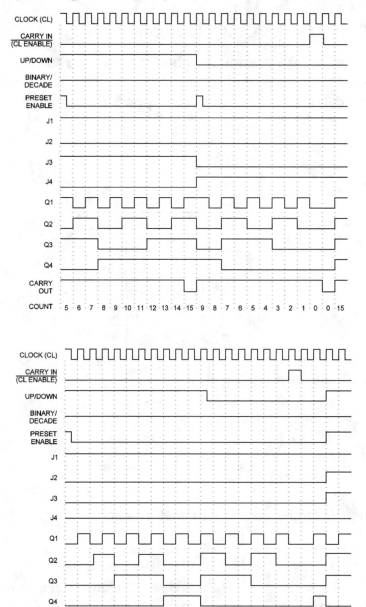

[Figure 2.32]

Functional Diagrams and Information for Designers

Operation Mode:

- LD and EN are grounded.
- With D/B with a "0" logic level, the circuit counts by tens. If in the "1" logic level the circuit counts by sixteens.
- If UP/DOWN is held at the "1" logic level, the circuit operates as an up counter and with the pin at the "0" logic level the circuit operates as a down counter.
- The changes in the count occur with the positive transition of the clock.
- CO is "0" for count 9 (bcd) and 15 (binary).
- Em ground stops the count.

Electrical Characteristics:

Characteristic	Conditions (Vdd)	Value	Units
Drain/Source Current (typ)	5V	0.88	mA
	10V	2.25	mA
	15V	8.8	mA
Maximum Clock Frequency (typ)	5V	3.1	MHz
	10V	7.4	MHz
	15V	9	MHz
Quiescent Device Current (max)	5V	5	µA
	10V	10	µA
	15V	20	µA
Supply Voltage Range	3V to 15V		V

Applications:

- Programmable Binary/Decade Counting
- A/D and D/A Conversion
- Up/Down Binary/Decade Counting
- Magnitude and Sign Generators
- Difference Counting

Observations:

- The up-down control can be changed only when the clock is at the "1" logic level.
- Cascading: Connecting the OUT pin of the first decade to EN of the second decade, it is possible to drive the circuit from a common clock for fully synchronous operation.

4030

Quad Exclusive-OR Gate

Description: This device is formed by four independent Exclusive-OR Gates.

Functional Diagram or/and Package:

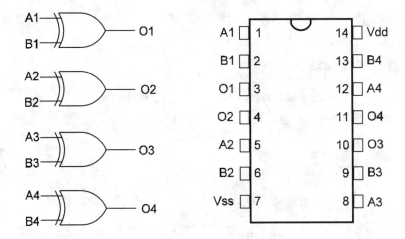

(Figure 2.33)

Pin Names:

Vdd – Positive Supply Voltage (3V to 15V)
Vss – Ground
A1, B1, A2, B2, A3, B3, A4, B4 – Inputs
O1, O2, O3, O4 – Outputs

Functional Diagrams and Information for Designers

Truth Table:

A	B	O
0	0	0
0	1	1
1	0	1
1	1	0

Operation Mode:

- The signals are applied to the inputs A and B.
- The output depends on the input signals (see the truth table).

Electrical Characteristics:

Characteristic	Conditions (Vdd)	Value (typ)	Units
Drain/Source Current (typ)	5V	1.2/0.65	mA
	10V	2.4/1.3	mA
	15V	-	-
Propagation Delay Time (typ)	5V	100	ns
	10V	40	ns
	15V	-	-
Quiescent Device Current (max)	5V	0.005	µA
	10V	0.001	µA
	15V	-	µA
Supply Voltage Range	3V to 15V		V

Other Devices:

The 4070 and 4507 are equivalent devices.

Applications:

- Even- and Odd-Parity Generators/Checkers
- Logic Functions
- Adders/Subtractors
- Logic Comparators

Observations:

Some versions became erratic due low input impedance.

4031

64-Stage Static Shift Register

Description: This package contains an integrated 64-stage fully static shift register. The circuit has two data inputs, DATA IN and Recirculate IN, as well as a MODE CONTROL input.

Functional Diagram and/or Package:

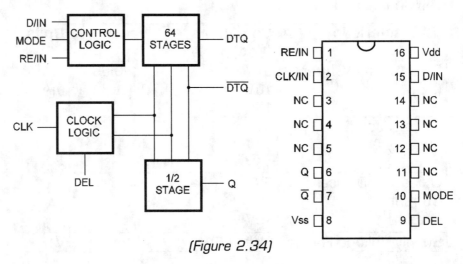

(Figure 2.34)

Pin Names:

Vdd – Positive Supply Voltage (3V to 15V)
Vss – Ground
RE/IN – Recirculate Input
CLK/IN – Clock Input
Q – Data Output
Q/ – Complementary Data Output
D/IN – Data In
MODE – Mode Control
DEL – Delayed Clock Output

Functional Diagrams and Information for Designers

Truth Table:

Mode Control (Data Selection)

MODE	D/IN	R/IN	Data into First Stage
0	0	X	0
0	1	X	1
1	X	0	0
1	X	1	1

Each Stage

Dn	CL	Qn
0	↑	0
1	↑	1
X	↓	NC

NC – No change
↓↑ - Level transitions
X – Don't care

Operation Mode:

a) MODE Input Grounded – mode 1
- MODE is grounded.
- Data to be stored is applied to the Data-In terminal (D/IN). Data enters the circuit at the positive transition of the clock signal.
- After 64 clock pulses, data applied to the input appears at pin 6 and the complement at pin 7.

b) MODE Input at "1" logic level – mode 2
- MODE input is at "1" logic level.
- Data is applied to the RE/IN Input and stored in the circuit with the positive transition of the clock signal.
- For recirculation, REC IN is connected to OUT.

Electrical Characteristics:

Characteristic	Conditions (Vdd)	Value	Units
Drain/Source Current (typ)	5V	0.88	mA
	10V	2.25	mA
	15V	8.8	mA
Maximum Clock Frequency (typ)	5V	3.2	MHz
	10V	8.0	MHz
	15V	10	MHz
Quiescent Device Current (max)	5V	20	µA
	10V	40	µA
	15V	80	µA
Supply Voltage Range	3V to 15V		V

Applications:

- Shift Registers (serial)
- Time Delay Applications

Observations:

- The terminal OUT can be used as an input to a following register. This allows for cascading stages in multiples of 64.
- The outputs can drive one TTL input.
- The clock input has a high input capacitance of about 60 pF.

4034

8-Stage TRI-STATE Bidirectional Parallel/Serial/Output Bus Register

Description: This circuit allows you to transfer 8-bit words between two bus systems and may also function as a universal 8-bit shift register.

Functional Diagram and/or Package:

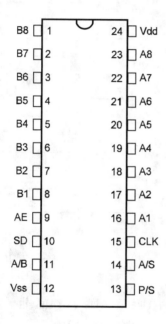

(Figure 2.35)

Pin Names:

Vdd – Positive Supply Voltage (3V to 15V)

Vss – Ground

AE – Tri-State Float

SD – Serial Data

A1 to A8 – Outputs for Buses

B1 to B8 – Outputs for buses

CLK – Clock

A/S – Asynchronous or Synchronous Mode Selection

P/S – Parallel/Serial Mode Selection

A/B – Line Selection

CMOS Sourcebook

Timing Diagram:
Figure 2.36 shows the time diagram for this device.

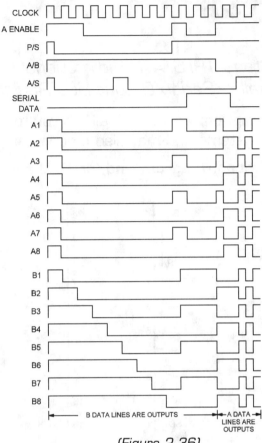

(Figure 2.36)

Operation Mode:
- A "1" applied to the P/S input allows data transfer into the register via the parallel data lines.
- The signals are transferred with the positive transition of the clock signal.

The control lines determine what the device is going to do:

- If A/B line selection is "1" then the eight A-bus lines operate as inputs and the eight B-bus lines act as outputs.
- If A/B line selection is "0" then the eight B-lines are now inputs and the eight A-bus lines act as outputs.
- The AE inputs are a feature that allows many registers to feed data to a common bus. The A date lines are enabled only when this input is "1".
- Data storage through recirculation of data in each register is accomplished by making the A/B input "1" and AE "0".
- When the Asynchronous/Synchronous (A/S) line is "1", transfer

Functional Diagrams and Information for Designers

occurs immediately. If "0", the transfer occurs with the positive transition of the clock input.

- If the P/S line is "1" the circuit is in the parallel data load mode. If in the "0" logic level, data is serially loaded. When serially loaded, data is transferred from stage to stage with the positive transition of the clock signal.
- P/S line must be "0" when P/S is "1" because asynchronous serial operation is not possible with this device.

Electrical Characteristics:

Characteristic	Conditions (Vdd)	Value	Units
Drain/Source Current (min)	5V	0.51	mA
	10V	1.3	mA
	15V	3.4	mA
Maximum Clock Frequency (typ)	5V	4	MHz
	10V	10	MHz
	15V	14	MHz
Quiescent Device Current (max)	5V	5	µA
	10V	10	µA
	15V	20	µA
Supply Voltage Range	3V to 15V		V

Applications:

- General Purpose Register (PIPO, SIPO, SISO)
- Phase and Frequency Comparators
- Double Bus Register Systems
- Address and Buffer Registers
- Shift Right/Shift Left Registers (with parallel loading)
- Johnson Ring Counters
- Pseudo-Random Code Generators
- Sample and Hold Registers

Observations:

Register expansion can be accomplished by simply cascading devices.

4035

4-Bit Parallel-in/Parallel-Out Shift Register

Description: This package contains a 4-stage Parallel-In/Parallel-Out (PIPO) Shift Register.

Functional Diagram and/or Package:

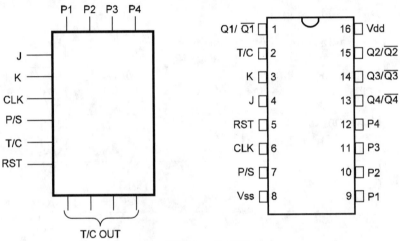

[Figure 2.37]

Pin Names:

Vdd – Positive Supply Voltage (3V to 15V)
Vss – Ground
Q1/Q1/, Q2/Q2/, Q3/Q3/, Q4/Q4/ – Outputs
RST – Reset
CLK – Clock
T/C – True Complement Input
P/S – Parallel/Serial Selection
J – Input Logic
K – Input Logic
P1, P2, P3, P4 – Parallel Inputs

Functional Diagrams and Information for Designers

Truth Table:

CL	J	K	R	Qn-1	Qn
↑	0	X	0	0	0
↑	1	X	0	0	1
↑	X	0	0	1	0
↑	1	0	0	Qn-1	Qn-1\ (toggle mode)
↑	X	1	0	1	1
↓	X	X	0	Qn-1	Qn-1
X	X	X	1	X	0

X – Don't care
↓↑ -Logic level transitions

Operation Mode:

a) Parallel Load Data
- Data is applied to the inputs P1 to P4. P1 is nearest the input.
- Next, LD must go the "1" logic level.
- In the next positive transition of the clock signal, data is loaded into the shift register.
- LD must be kept in the "1" level until after clocking, and then it can drop.

b) Serial In/Serial Out or Serial In/Parallel Out
- J and K are connected together.
- TC, RST, and LD are also tied together.
- Input data is applied to the J and K inputs. Data will appear at the output Q1 after a positive transition of the clock signal.
- The second clock pulse transfers the signal to the next stage and so on.

Electrical Characteristics:

Characteristic	Conditions (Vdd)	Value	Units
Drain/Source Current (typ)	5V	0.88	mA
	10V	2.25	mA
	15V	8.8	mA
Maximum Clock Frequency (typ)	5V	2.5	MHz
	10V	6	MHz
	15V	9	MHz
Quiescent Device Current (max)	5V	0.3	µA
	10V	0.5	µA
	15V	1.0	µA
Supply Voltage Range	3V to 15V		V

Applications:

- Counters
- Registers
- Serial-to-Parallel and Parallel-to-Serial Converters
- Shift/Shift Right Registers
- Control Applications
- Code Conversion
- Sequence Generation

Observations:

The words applied to the inputs must arrive least significant bit first. A sign bit must follow the most significant bit.

4041

Quad True/Complement Buffer

Description: This package contains four true/complement buffers consisting of N and P channel transistor having low channel resistance and high current capability.

Functional Diagram and/or Package:

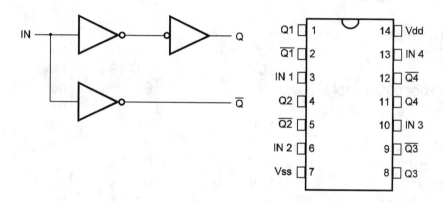

(Figure 2.38)

Pin Names:

Vdd – Positive Supply Voltage (3V to 15V)
Vss – Ground
IN1, IN2, IN3, IN4 – Inputs
Q1, Q2, Q3, Q4 – True Outputs
Q1/, Q2/, Q3/, Q4/ – Complement Outputs

Truth Table:

IN	Q	Q/
0	1	0
1	0	1

Operation Mode:

- The output of one buffer/inverter is applied to the input of the other buffer.
- The logic level applied to the input of the block appears inverted in the first output and inverted again (true) in the second output.
- A "1" in the input causes to a "0" to appear in the first output and a "1" in the second output.

Electrical Characteristics:

Characteristic	Conditions (Vdd)	Value	Units
Drain/Source Current (typ)	5V	1.5/2.8	mA
	10V	4.0/8.0	mA
	15V	9.0/18	mA
Propagation Delay Time (True Output) – (typ)	5V	60	ns
	10V	35	ns
	15V	25	ns
Quiescent Device Current (max)	5V	0.01	µA
	10V	0.01	µA
	15V	0.01	µA
Supply Voltage Range	3V to 15V		V

Other Devices:

The 4069 and 4050 can be used in the same function.

Applications:

- High Current Driver (sink and source)
- CMOS-to-TTL Driver
- Display Driver
- Digital Amplifier
- Transmission Line Driver

Observations:

This device is TTL compatible.

4042

Quad Clocked D Latch

Description: This device is formed by four D-type latches with level-controlled storage command pins.

Functional Diagram and/or Package:

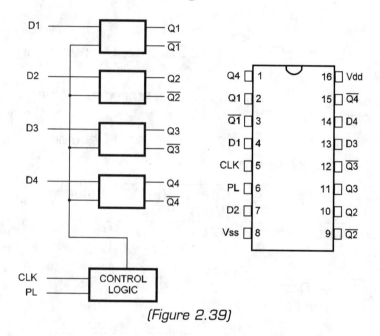

(Figure 2.39)

Pin Names:

Vdd – Positive Supply Voltage (3V to 15V)
Vss – Ground
D1, D2, D3, D4 – D Inputs
Q1, Q2, Q3, Q4 – Outputs
Q1/, Q2/, Q3/, Q4/ – Complementary Outputs
CLK – Clock
PL – Polarity

Truth Table:

CLK	Polarity	Q
0	0	Q
↑	0	Latch
1	1	D
↓	1	Latch

↓↑- Logic level changes

Operation Mode:
- With PL=0 and ST=0, data applied to D appears in the outputs (Q and Q/).
- Passing PL to "1", data in the input is stored at the negative transition of the clock pulse.
- The output follows one state of the Store Control.

Electrical Characteristics:

Characteristic	Conditions (Vdd)	Value	Units
Drain/Source Current (typ)	5V	0.88	mA
	10V	2.25	mA
	15V	8.8	mA
Propagation Delay Time (Data in Q) – (typ)	5V	175	ns
	10V	75	ns
	15V	60	ns
Quiescent Device Current (max)	5V	1	µA
	10V	2	µA
	15V	4	µA
Supply Voltage Range	3V to 15V		V

Applications:
- Buffer Storage • Digital Logic • Holding Registers

Observations:
Stages using this device may not be cascaded.

Functional Diagrams and Information for Designers

4043

Quad Tri-State NOR R/S Latches

Description: Four independent R-S flip-flops are found in this package. The device has a common tri-state enable control for the four flip-flops.

Functional Diagram and/or Package:

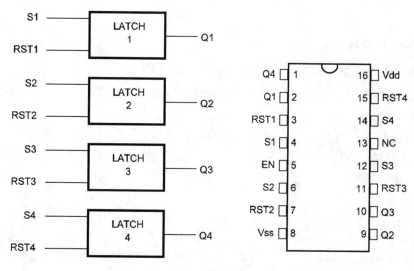

(Figure 2.40)

Pin Names:

  Vdd – Positive Supply Voltage (3V to 15V)
  Vss – Ground
  S1, S2, S3, S4 – Set
  RST1, RST2, RST3, RST4 – Reset
  Q1, Q2, Q3, Q4 – Outputs
  EN – Enable
  NC – Not Connected

Truth Table:

S	RST	EN	Q
X	X	0	OC
0	0	1	NC
1	0	1	1
0	1	1	0
1	1	1	(*)

X – Don't care
OC – Tri-state
NC – No change
(*) Dominated by S=1 input

Operation Mode:

- In normal operation, S and RST are "0".
- When S and RST are "1", the output goes to "1". This is a disallowed state.
- When EN is "0", the output goes to the third state (high-impedance state). With the input EN=1, the outputs are connected to the internal circuit.

Electrical Characteristics:

Characteristic	Conditions (Vdd)	Value	Units
Drain/Source Current (typ)	5V	1.0/0.4	mA
	10V	2.6/1.0	mA
	15V	6.8/3.0	mA
Propagation Delay S or R to Q (typ)	5V	175	Ns
	10V	75	ns
	15V	60	ns
Quiescent Device Current (max)	5V	1.5	µA
	10V	3.0	µA
	15V	4.0	µA
Supply Voltage Range	3V to 15V		V

Functional Diagrams and Information for Designers

Other Devices:
- The 4044 is similar, but it has NAND inputs.

Applications:
- Strobed Registers
- Digital Logic
- Holding Registers (for Multimode Register Systems)
- Four Bits of Independent Storage

Observations:

Stages using this device may not be cascaded and are not suitable for counters or shift registers.

4044

Quad Tri-State NAND R/S Latches

Description: This package contains four independent NAND R/S latches with tri-state outputs. The third state is controlled by a common EN pin.

Functional Diagram and/or Package:

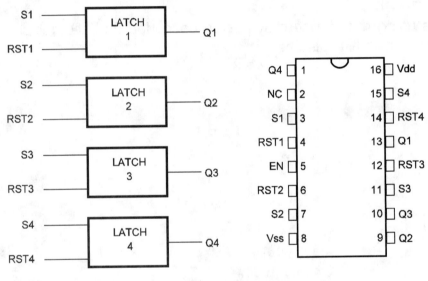

(Figure 2.41)

Pin Names:

Vdd – Positive Supply Voltage (3V to 15V)
Vss – Ground
EN – Enable
S1, S2, S3, S4 – Set
RST1, RST2, RST3, RST4 – Reset
Q1, Q2, Q3, Q4 – Outputs
NC – Not Connected

Functional Diagrams and Information for Designers

Truth Table:

S	RST	EN	Q
X	X	0	OC
1	1	1	NC
0	1	1	1
1	0	1	0
0	0	1	(*)

OC – Tri-state
NC – No change
(*) – Dominated by R=0 input

Operation Mode:
- S and RST are normally "1".
- When S goes to "0", Q passes to "1" and maintains the logic level.
- When RST goes to "0", Q passes to "0" and stays in this state.
- S and RST can't be "0" at the same time because both outputs pass to "0". This is a disallowed state.
- EN controls the tri-state function. The outputs go to the third state when EN=0.

Electrical Characteristics:

Characteristic	Conditions (Vdd)	Value	Units
Drain/Source Current (typ)	5V	1.0/0.4	mA
	10V	2.6/1.0	mA
	15V	6.8/3.0	mA
Propagation Delay S or R to Q (typ)	5V	175	ns
	10V	75	ns
	15V	60	ns
Quiescent Device Current (max)	5V	5	µA
	10V	10	µA
	15V	20	µA
Supply Voltage Range	3V to 15V		V

Other Devices:
- The 4043 is similar, but it has NOR inputs.

Applications:
- Holding Registers (for Multiregister Systems)
- Strobed Registers
- Four Bits of Independent Storage
- Digital Logic

Observations:

Do not cascade this device or use as shift register.

4046

Micropower Phase-Locked Loop (PLL)

Description: This package contains a complete PLL with a low-power Voltage Controlled Oscillator (VCO) and two Phase Comparators that have a common signal input amplifier and a common comparator input. The device also contains a 5.2V zener diode for supply regulation, if necessary.

Functional Diagram and/or Package:

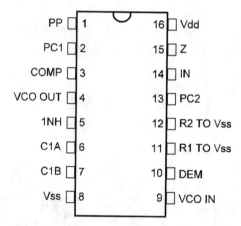

(Figure 2.42)

Pin Names:

Vdd – Positive Supply Voltage (3V to 15V)
Vss – Ground
PP – Phase Pulses
PC – Phase Comparator Output
COMP – Comparator
VCO OUT – VCO Output
INH – Inhibit

C1A, C1B – Capacitor C1
Z – Zener
IN – Signal In
R1, R2 – Resistors R1 and R2 Connections
DEM – Demodulator Output
VCO IN – VCO Input

Truth Table: none

CMOS Sourcebook

Operation Mode:

a) VCO Frequency

- The frequency is determined by the voltage applied to pin 9, the capacitor between pin 6 and 7, and the resistor to pin 11 (Figure 2.43).
- The capacitor has a minimum value of 50 pF, and the resistor on pin 12 can assume values between 10k and infinity.
- Output is in pin 4.

b) Phase detector #1

- It is an exclusive OR system.
- This system must have square waves on both pins 3 and 14.
- This system is applied to narrow frequency ranges.

c) Logic Frequency/Phase detector #2

- Frequency range of 1000:1.
- Any input duty cycle.

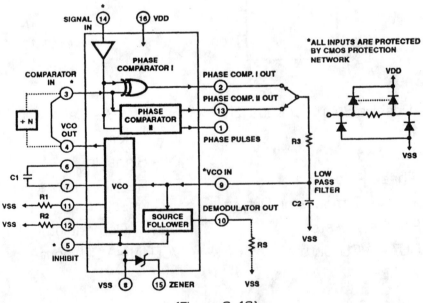

(Figure 2.43)

Functional Diagrams and Information for Designers

Electrical Characteristics:

Characteristic	Conditions (Vdd)	Value	Units
Drain/Source Current (max)	5 V	0.88	mA
	10 V	2.25	mA
	15 V	8.8	mA
Maximum Operating Frequency (typ)	5 V	0.8	MHz
	10 V	1.2	MHz
	15 V	1.6	MHz
Quiescent Device Current (max)	5 V	0.005	µA
	10 V	0.01	µA
	15 V	0.015	µA
Supply Voltage Range	3 to 15		V
Input Capacitance	7.5		PF

Other Devices:
- The LM/NE567 is a PLL that can be used in many applications where the 4046 is found, although it isn't a CMOS device.

Applications:
- FM Demodulator and Modulator
- Tone Decoders
- Frequency Synthesis
- Data Synchronization
- Voltage-to-Frequency Converters
- Modems (FSK)
- Signal Conditioning
- Tone Generators

Observations:
Practical circuits using this device are covered in Part 3.

4047

Monostable/Astable Multivibrator

Description: This package contains a gateable astable multivibrator with resources to permit positive or negative edge-triggered monostable multivibrator action with re-triggering and external counting options.

Functional Diagram and/or Package:

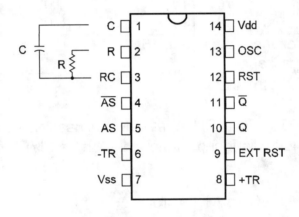

(Figure 2.44)

Pin Names:

Vdd – Positive Supply Voltage (3V to 15V)
Vss – Ground
C – Timing Capacitor
R – Timing Resistor
RC – RC Common
AS – Astable Input

AS/ – Astable Complementary Input
-TR, +TR – Trigger
OSC – Oscillator Output
RTR – Retrigger
Q – Output
Q/ – Complementary Output
EX RST – External Reset

Functional Diagrams and Information for Designers

Truth Table:

Function	Terminal Connections			Output Pulse from	Pulse Width
	To Vdd	To Vss	Input Pulse to		
Astable Multivibrator					
Free running:	4,5,6,14	7,8,9,12		10,11,13	ta(10,11)=4.40RC
True Gating:	4,6,14	7,8,9,12	5	10,11,13	
Complement Gating:	6,14	5,7,8,9,12	4	10,11,13	ta(13) = 2.20RC
Monostable Multivibrator:					
Positive Edge trigger:	4,14	5,6,7,9,12	8	10,11	
Negative Edge Trigger:	4,8,14	5,7,9,12	6	10,11	Ta(10,11)=2,48RC
Retriggerable:	4,14	5,6,7,9	8,12	10,11	
External Countdown:	14	5,6,7,8,9,12	(*)	(*)	

(*) Input pulses are applied to RS. OUT connected to AS/.

Operation Mode:

a) Astable Operation
  - AS/ is put at the "1" logic level or an AS put at the "0" logic level (or both).
  - The period of the output signal is a function of RC.

b) Monostable Operation
  - The pulses applied to the AS/ input can be used to gate the multivibrator.
  - The device triggers with the positive-going pulse with the TR input.
  - Retrigger is obtained using the RTR input.

c) Countdown option:
  - Couple an external counter to Q.
  - A pulse applied to the EXT RST input may terminate the output pulse at any time for monostable operation (see truth table for more details).

Electrical Characteristics:

Characteristic	Conditions (Vdd)	Value	Units
Drain/Source Current (typ)	5V	0.88	mA
	10V	2.25	mA
	15V	8.8	mA
Propagation Delay Time (astable to OSC OUT) – (max)	5V	200	ns
	10V	199	ns
	15V	80	ns
Quiescent Device Current (max)	5V	5	µA
	10V	10	µA
	15V	20	µA
Supply Voltage Range	3V to 15V		V

Other Devices:
- In many applications, the popular 555 and the CMOS equivalent 7555 can replace the 4047.

Applications:
- Frequency Division/Multiplication
- Timing Applications
- Envelope Detection
- Time Delay Circuits
- Frequency Discriminators
- Missing Pulse Detectors

Observations:
The astable configuration produces 50 percent duty-cycle square wave signals.

Functional Diagrams and Information for Designers

4048

Tri-State Expandable 8-Function 8-Input Gate

Description: This package contains a set of gates that can be programmed to perform any common 8-input function or combined 4-input function. The 8-input functions are, AND/NAND-OR/NOR. The combined two 4-input functions are, OR/AND, OR/NAND, AND-OR, AND/NOR.

Functional Diagram and/or Package:

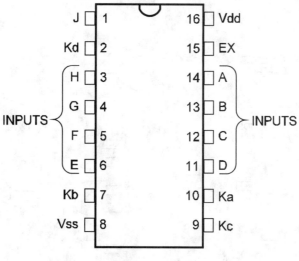

(Figure 2.45)

Pin Names:

Vdd – Positive Supply Voltage (3V to 15V)
Vss – Ground
A, B, C, D, E, F, G, H – Inputs
Ka, Kb, Kc – Control Lines
Kd – Tri-state Control
EX – Expand
J – Output

CMOS Sourcebook

Truth Table:

Output Function	Ka	Kb	Kc	Unused Inputs
NOR	0	0	0	Vss
OR	0	0	1	Vss
OR/AND	0	1	0	Vss
OR/NAND	0	1	1	Vss
AND	1	0	0	Vdd
NAND	1	0	1	Vdd
AND/NOR	1	1	0	Vdd
AND-OR	1	1	1	Vdd

Kd = 1 – Normal Inverter Action

Kd = 0 – High Impedance (tri-state)

See Figure 2.46 for the logic configurations.

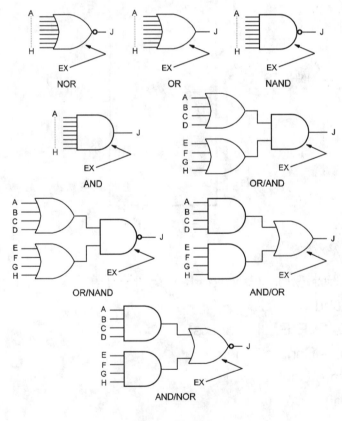

(Figure 2.46)

Functional Diagrams and Information for Designers

Operation Mode:
- The logic levels applied to the control lines Ka, Kb, and Kc determine the function performed by the device (see the truth table).
- Kd determines the tri-state of the output. When in the "0" logic level, the output goes to a high-impedance state.
- The EX (Expand) input is used to increase the number of inputs. Two 4048 can be cascaded by providing a 16-input multifunction gate.

Electrical Characteristics:

Characteristic	Conditions (Vdd)	Value (typ)	Units
Drain/Source Current	5V	4.0	mA
	10V	11	mA
	15V	23	mA
Propagation Delay Time	5V	425	ns
	10V	200	ns
	15V	180	ns
Quiescent Device Current (max)	5V	0.01	µA
	10V	0.01	µA
	15V	0,01	µA
Supply Voltage Range	3V to 15V		V

Other Devices:
- Devices with specific functions can be found in this list.

Applications:
- Logic Function Selector
- General Purpose Logic (Encoding/Decoding)
- Digital Control of Logic

Observations:
When Expand is not used it must be connected to Vss.

4049

Hex Inverting Buffer

Description: This device contains 6 independent inverting buffers.

Functional Diagram and/or Package:

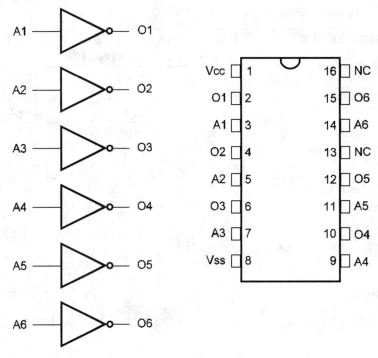

(Figure 2.47)

Pin Names:

Vcc – Positive Supply Voltage (3V to 15V)
Vss – Ground
A1, A2, A3, A4, A5, A6 – Inputs
O1, O2, O3, O4, O5, O6 – Outputs
NC – Not Connected

Functional Diagrams and Information for Designers

Truth Table:

A	O
0	1
1	0

Operation Mode:
- Logic levels applied to the inputs appear inverted in the outputs (See truth table).

Electrical Characteristics:

Characteristic	Conditions (Vdd)	Value	Units
Drain/Source Current	5V	5/1.6	mA
	10V	12/3.6	mA
	15V	40/12	mA
Propagation Delay Time	5V	30	ns
	10V	20	ns
	15V	15	ns
Quiescent Device Current (max)	5V	1	µA
	10V	2	µA
	15V	4	µA
Supply Voltage Range	3V to 15V		V

Applications:
- CMOS to TTL Drivers
- Digital Amplifiers (High Current Sink/Source)
- High-to-Low Logic Converters

Observations:
- TTL compatible when powered from 5V supplies.
- Observe the unusual pinout for this device.

4050

Hex Non-Inverting Buffer

Description: The circuit inside this package is formed by six independent non-inverting buffers.

Functional Diagram and/or Package:

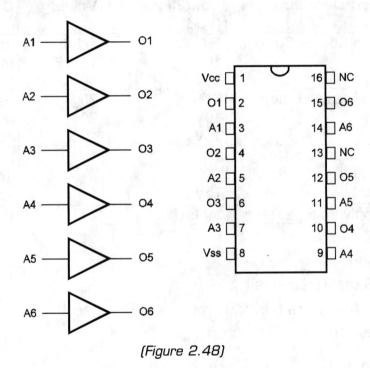

(Figure 2.48)

Pin Names:

Vcc – Positive Supply Voltage (3V to 15V)
Vss – Ground
A1, A2, A3, A4, A5, A6 – Inputs
O1, O2, O3, O4, O5, O6 – Outputs
NC – Not Connected

Functional Diagrams and Information for Designers

Truth Table:

A	O
0	0
1	1

Operation Mode:

The output voltage swing is determined by the voltage applied to pin 1.

A "1" input applied to each buffer provides a "1" output.

The output can drive two Regular TTL gates or four LS TTL gates.

Electrical Characteristics:

Characteristic	Conditions (Vdd)	Value	Units
Drain/Source Current (typ)	5V	5/1.6	mA
	10V	12/3.6	mA
	15V	40/12	mA
Propagation Delay Time (typ)	5V	60	ns
	10V	25	ns
	15V	20	ns
Quiescent Device Current (max)	5V	1	µA
	10V	2	µA
	15V	4	µA
Supply Voltage Range	3 to 15		V

Other Devices:

Two inverting buffers such as the 4049 can be used to perform a non-inverting buffer.

Applications:

CMOS-to-TTL Interfacing sourcing 5V to pin 1

Digital Amplifier (High-Current Source/Sink)

Observations:

Unusual connections are used in this package

4051

Single 8-Channel Analog Multiplexer/Demultiplexer (Mux-Demux)

Description: This package contains a set of digitally controlled switches with low ON impedance and very high OFF leakage currents. Analog signals can be controlled by digital inputs using this device.

Functional Diagram and/or Package:

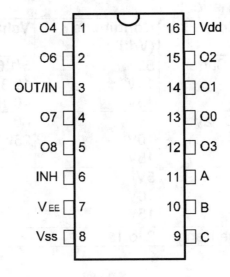

(Figure 2.49)

Pin Names:

Vdd – Positive Supply Voltage (3V to 15V)
Vss – Ground
I/O0, I/O1, I/O2, I/O3, I/O4, I/O5, I/O6, I/O7 – Inputs or Outputs
(I/O) OUT/IN – Input Selection
A, B, C – Control Lines
INH – Inhibit Vee = -5V

Functional Diagrams and Information for Designers

Truth Table:

INPUT STATES				"ON" CHANNELS
INH	A	B	C	OUT (pin 3)
0	0	0	0	0
0	0	0	1	1
0	0	1	0	2
0	0	1	1	3
0	1	0	0	4
0	1	0	1	5
0	1	1	0	6
0	1	1	1	7
1	X	X	X	None

X - Don't care

Operation Mode:

a) Analog Mode

- +5V is applied to pin 16 and -5V to pin 7 – Pin 8 remains grounded.
- The logic levels in the A, B, and C control lines determine the selected channel. INH must be "0".
- If INH = 1, no channel is selected.
- The maximum amplitude of the analog signal can't exceed 10Vpp.

b) Digital Mode

- Pin 7 is grounded. Pin 16 is sourced with voltages between 3V and 15V and also is grounded.
- The logic levels applied to the control lines (A,B, and C) determine the selected channel. In the selection, INH must be "0".
- If INH =1 no channel is selected.

In the OFF state the device is an open circuit, and in the ON state it is a low-resistance resistor (see Electrical Characteristics).

Electrical Characteristics:

Characteristic	Conditions (Vdd)	Value	Units
On Resistance (typ)	5V	1000	Ω
	10V	400	Ω
	15V	240	Ω
Propagation Delay Time from Address to Signal Output (typ)	5V	500	ns
	10V	180	ns
	15V	120	ns
Quiescent Device Current (max)	5V	5	µA
	10V	10	µA
	15V	20	µA
Supply Voltage Range	3 to 15		V

Other Devices:
- The 4052 is similar but a dual 1-of-4 Mux/Demux. The 4053 is also similar, but with a triple 1-of-2 configuration.

Applications:
- Analog and Digital Multiplexing and Demultiplexing
- Signal Gating
- A/D and D/A Converters

Observations:
- The load resistance must not be lower than 100 ohms.
- The higher current to be controlled by the circuit must be limited to 25 mA.

4052

Dual 4-Channel Analog Multiplexer/Demultiplexer (Mux-Demux)

Description: This package contains two 4-Channel Mux/Demux with two binary control lines and an inhibit input. The input signals may be used to select one of the 4 pairs of channels to be turned ON and connected from the differential analog inputs to the differential outputs.

Functional Diagram and/or Package:

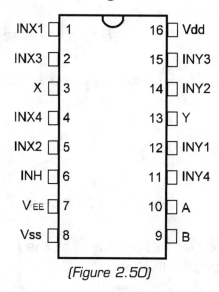

(Figure 2.50)

Pin Names:

Vdd – Positive Supply Voltage (3V to 15V)
Vss – Ground
INX1, INX2, INX3, INX4, INY1, INY2, INY3, INY4 – Inputs/Outputs
X, Y – Outputs/Inputs
INH – Inhibit
Vee – -5V
A, B – Control Inputs

Truth Table:

Input States			On Channels
INH	B	A	
0	0	0	0X, 0U
0	0	1	1X, 1Y
0	1	0	2X, 2Y
0	1	1	3X, 3Y

Operation Mode:

a) Analog Mode
- -5V is sourced to pin 7 – Pin 16 is connected to a +5V source and pin 8 to ground.
- The control inputs (A, B) determine the selected channel according to the truth table.
- INH must be "0".
- If INH = 1, no channel is selected.

b) Digital Mode
- Pin 7 and 8 are grounded and Vdd between 3 and 15V is sourced to pin 16.
- The selected channel is determined by the logic level applied to the Control Lines A and B. INH is "0".
- If INH = 1 no channel is selected.

In the OFF state the device is an open circuit, and when ON, represents a very low resistor (see Electrical Characteristics).

Functional Diagrams and Information for Designers

Electrical Characteristics:

Characteristic	Conditions (Vdd)	Value (typ)	Units
On Resistance (typ)	5V	1000	Ω
	10V	400	Ω
	15V	250	Ω
Propagation Delay Time from Address to Signal Output (typ)	5V	500	ns
	10V	180	ns
	15V	120	ns
Quiescent Device Current (max)	5V	5	μA
	10V	10	μA
	15V	20	μA
Supply Voltage Range	3 to 15		V

Other Devices:
- The 4051 is a 1-of-8 switch and the 4053 is a triple 1-of-2 switch.

Applications:
- Analog and Digital Multiplexing and Demultiplexing
- A/D and D/A Converters
- Signal Gating

Observations:
- The minimum load resistance is 100 ohm.
- The maximum controlled current is 25 mA.
- The maximum amplitude of analog signals is 10Vpp.

4053

Triple 2-Channel Analog Multiplexer/Demultiplexer (Mux-Demux)

Description: This package contains 3 differential 2-channel analog Mux/Demux. This device has three separated control inputs and an inhibit input. Each control input selects one of a pair of channels that are connected in a single pole-double through configuration.

Functional Diagram or/and Package:

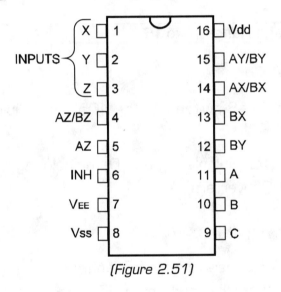

(Figure 2.51)

Pin Names:

   Vdd – Positive Supply Voltage (3V to 15V)
   Vss – Ground
   AX, BX, AY, BY, AZ, BZ – Inputs/Outputs
   X, Y, Z – Outputs/Inputs
   INH – Inhibit
   Vee – -5V
   A, B, C – Control Inputs

Functional Diagrams and Information for Designers

Truth Table:

INH	A, B or C	
0	0	AX, BX or CX
0	1	AY, BY or CY
1	X	None

X – Don't care

Operation Mode:

a) Analog Mode
- Apply -5V to pin 7 and ground pin 8. Pin 16 is sourced with +5V.
- The selected channel is determined by the logic levels in the control lines A, B, and C. INH must be "0".
- A controls the first switch. If A is high, AX is connected to X. If A is low, AY is connected to X. See truth table for more details.
- If INH =1 all the switches are OFF and no channel is selected.

b) Digital Mode
- Ground pin 7 and pin 8. Pin 16 is sourced with voltages between 3V and 15V.
- The selected channel of each switch is determined by the logic levels applied to the control lines A, B, and C. INH must be "0".
- If INH = 1, no channel is selected.

In the ON state the devices acts as a low value resistor (see Electrical Characteristics).

Electrical Characteristics:

Characteristic	Conditions (Vdd)	Value	Units
On Resistance (typ)	5V	1050	mA
	10V	400	mA
	15V	240	mA
Propagation Delay Time from Address to Signal Outputs (typ)	5V	500	ns
	10V	180	ns
	15V	120	ns
Quiescent Device Current (max)	5V	5	µA
	10V	10	µA
	15V	20	µA
Supply Voltage Range	3 to 15		V

Other Devices:
- The 4051 is a 1-of-8 switch and the 4052 is a dual 1-of-4 switch. Each has the same electrical characteristics and operation mode.

Applications:
- Analog and Digital Multiplexing and Demultiplexing
- Signal Gating
- A/D and D/A Converters

Observations:
- Maximum amplitude of the analog signal must be limited to 10Vpp.
- The minimum load resistance is 100 ohm.
- The maximum current through the device is 25 mA.

4066

Quad Bilateral Switch

Description: This package contains four independent analog/digital bilateral switches. The switches may be used for the multiplexing of signals. This circuit is equivalent to the 4016 but with lower resistance in the ON state.

Functional Diagram and/or Package:

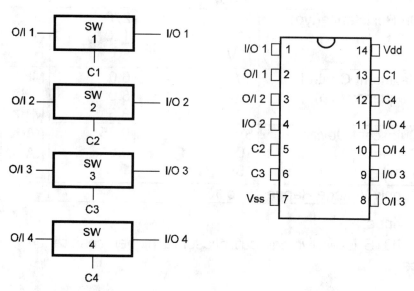

(Figure 2.52)

Pin Names:

Vdd – Positive Supply Voltage
Vss – Ground
I/O1, O/I1, I/O2, O/I2, I/O3, O/I3, I/O4, O/I4 – Inputs/Outputs
C1, C2, C3, C4 – Control Inputs

Operation Mode:
- All the switches are independent.
- When operating with digital signals, Vdd is sourced to pin 14 (3V to 15V) and pin 7 is grounded.
- When operating with analog signals, pin 14 is sourced with +5V and pin 7 with -5V. The amplitude of analog signals must be limited to 10 Vpp.
- The switches are ON when the control input is "1", and OFF when it is "0".

Electrical Characteristics:

Characteristic	Conditions (Vdd)	Value	Units
On Resistance (typ)	5V	270	Ω
	10V	120	Ω
	15V	80	Ω
Maximum Control Frequency (typ)	5V	6.0	MHz
	10V	8.0	MHz
	15V	8.5	MHz
Quiescent Device Current (max)	5V	0.25	µA
	10V	0.5	µA
	15V	1	µA
Supply Voltage Range	3 to 15		V

Other Devices:
- The 4016 is equivalent but presents higher resistance in the ON state.

Applications:
- Analog Signal Switching and Multiplexing
- Modulators and Demodulators
- Digital Signal Switching
- Digital Multiplexing
- A/D and D/A Converters
- Digital Control (Impedance, Phase, Frequency, Gain, etc.)
- Choppers

Observations:
External logic is needed if more than one switch is controlled by a circuit.

4069

Hex Inverter

Description: This device is formed by six independent inverters. It is a general-purpose device intended for applications where high noise immunity is required.

Functional Diagram and/or Package:

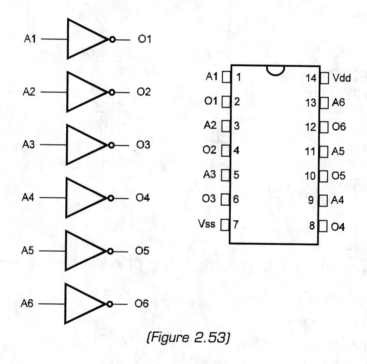

(Figure 2.53)

Pin Names:

Vdd – Positive Supply Voltage (3V to 15V)
Vss – Ground
A1, A2, A3, A4, A5, A6 – Inputs
O1, O2, O3, O4, O5, O6 – Outputs

Truth Table:

A	O
0	1
1	0

Operation Mode:
The logic signals applied to the input of each inverter appear inverted in the output (see Truth Table).

Electrical Characteristics:

Characteristic	Conditions (Vdd)	Value	Units
Drain/Source Current (typ)	5V	0.88	mA
	10V	2.25	mA
	15V	8.8	mA
Propagation Delay Time (typ)	5V	50	ns
	10V	30	ns
	15V	25	ns
Quiescent Device Current (max)	5V	0.25	µA
	10V	0.5	µA
	15V	1.0	µA
Supply Voltage Range	3 to 15		V

Other Devices:
- If you need more output current, the 4049 is recommended.

Applications:
- Inversion of Logic
- Oscillators
- Pulse Conditioning
- Digital Amplifiers (high impedance)

Observations:
This circuit can drive two low-power TTL inputs or one TTL LS input.

Functional Diagrams and Information for Designers

4070

Quad 2-Input Exclusive-OR Gate

Description: Four independent Exclusive-OR gates are found inside this package.

Functional Diagram or/and Package:

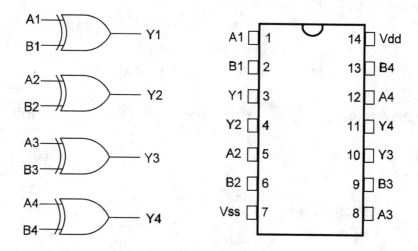

(Figure 2.54)

Pin Names:

Vdd – Positive Supply Voltage (3V to 15V)
Vss – Ground
A1, B1, A2, B2, A3, B3, A4, B4 – Inputs
Y1, Y2, Y3, Y4 – Outputs

Truth Table:

A	B	Y
0	0	0
0	1	1
1	0	1
1	1	0

Operation Mode:
- The gates can be used independently.
- The logic levels are applied to the inputs of each gate.
- The output goes to a logic level determined by the combination of the logic inputs according to the truth table.

Electrical Characteristics:

Characteristic	Conditions (Vdd)	Value	Units
Drain/Source Current (typ)	5V	0.88	mA
	10V	2.25	mA
	15V	8.8	mA
Propagation Delay Time (typ)	5V	110	ns
	10V	50	ns
	15V	40	ns
Quiescent Device Current (max)	5V	0.25	µA
	10V	0.5	µA
	15V	1.0	µA
Supply Voltage Range	3 to 15		V

Other Devices:
The 4508 is an equivalent device. The 4030 is also an equivalent device but is not used much anymore.

Applications:
- Logic Functions
- Parity Generators/Checkers
- Adders/Subtractors

Functional Diagrams and Information for Designers

4071

Quad 2-Input OR Buffered B Series Gate

Description: This package contains four independent buffered B series OR gates. The high-output capabilities of this device make it ideal for driving two low-power TTL inputs or one TTL LS.

Functional Diagram and/or Package:

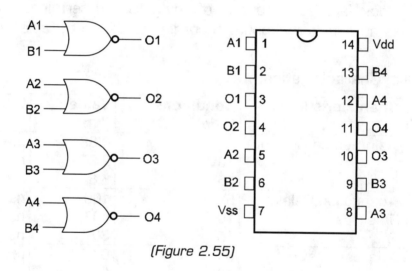

(Figure 2.55)

Pin Names:

Vdd – Positive Supply Voltage (3V to 15V)
Vss – Ground
A1, B1, A2, B2, A3, B3, A4, B4 – Inputs
O1, O2, O3, O4 – Outputs

Truth Table:

A	B	O
0	0	0
0	1	1
1	0	1
1	1	1

Operation Mode:
- The gates are independent.
- The logic level at the output of any gate is determined by the logic levels applied to the inputs according to the truth table.

Electrical Characteristics:

Characteristic	Conditions (Vdd)	Value (typ)	Units
Drain/Source Current	5V	0.88	mA
	10V	2.25	mA
	15V	8.8	mA
Propagation Delay Time	5V	100	ns
	10V	40	ns
	15V	30	ns
Quiescent Device Current (max)	5V	0.25	µA
	10V	0.5	µA
	15V	1.0	µA
Supply Voltage Range	3 to 15		V

Other Devices:
NOR gates can be implemented using OR gates and an inverter.

Applications:
- Logic Functions
- Oscillators

4072

Dual 4-Input OR Gate

Description: This package contains two independent 4-input OR gates.

Functional Diagram and/or Package:

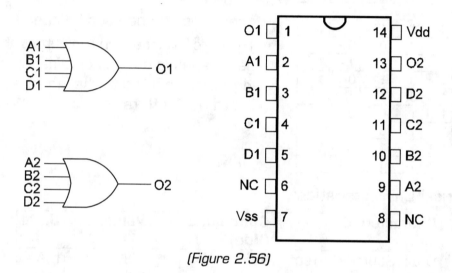

(Figure 2.56)

Pin Names:

Vdd – Positive Supply Voltage (3V to 15V)
Vss – Ground
A1, B1, C1, D1, A2, B2, C2, D2 – Inputs
O1, O2 – Outputs

Truth Table:

A	B	C	D	O
0	0	0	0	0
0	0	0	1	1
0	0	1	0	1
0	0	1	1	1
0	1	0	0	1
0	1	0	1	1
0	1	1	0	1
0	1	1	1	1
1	0	0	0	1
1	0	0	1	1
1	0	1	0	1
1	0	1	1	1
1	1	0	0	1
1	1	0	1	1
1	1	1	0	1
1	1	1	1	1

Operation Mode:

- The gates can be used independently.
- The logic level found in the output of each gate depends on the combination of the logic levels applied to the inputs according to the truth table.

Electrical Characteristics:

Characteristic	Conditions (Vdd)	Value	Units
Drain/Source Current (typ)	5V	0.88	mA
	10V	2.2	mA
	15V	8.0	mA
Propagation Delay Time (typ)	5V	125	ns
	10V	60	ns
	15V	45	ns
Quiescent Device Current (max)	5V	0.25	µA
	10V	0.5	µA
	15V	1	µA
Supply Voltage Range	3 to 15		V

Applications:

- Logic Functions
- Oscillators
- Digital Amplifier

4073

Double Buffered 3-Input AND Gate

Description: This device is formed by three independent 3-input AND gates in the same package.

Functional Diagram and/or Package:

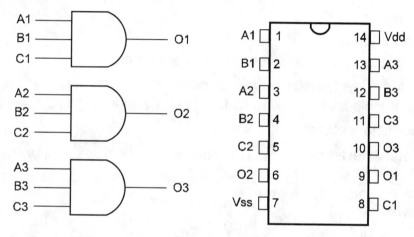

(Figure 2.57)

Pin Names:

Vdd – Positive Supply Voltage (3V to 15V)
Vss – Ground
A1, B1, C1, A2, B2, C2, A3, B3, C3 – Inputs
O1, O2, O3 – Outputs

Truth Table:

A	B	C	O
0	0	0	0
0	0	1	0
0	1	0	0
0	1	1	0
1	0	0	0
1	0	1	0
1	1	0	0
1	1	1	1

Operation Mode:
- The gates in this package are independent.
- The logic level at the output of each gate depends on the combination of logic levels applied to the inputs according to the truth table.

Electrical Characteristics:

Characteristic	Conditions (Vdd)	Value	Units
Drain/Source Current (typ)	5V	0.88	mA
	10V	2.2	mA
	15V	8.0	mA
Propagation Delay (typ)	5V	130	ns
	10V	60	ns
	15V	40	ns
Quiescent Device Current (max)	5V	0.25	µA
	10V	0.5	µA
	15V	1	µA
Supply Voltage Range	3 to 15		V

Applications:
- Logic Functions
- Oscillators
- Digital Amplifiers

Observations:
The output can drive two low-power TTL inputs or one TTL LS input.

Functional Diagrams and Information for Designers

4075

Triple 3-Input OR Gate

Description: This package contains three, buffered independent 3-input OR gates.

Functional Diagram and/or Package:

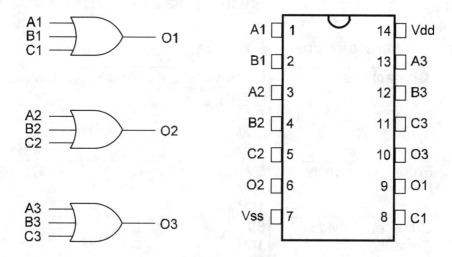

(Figure 2.58)

Pin Names:

Vdd – Positive Supply Voltage (3V to 15V)
Vss – Ground
A1, B1, C1, A2, B2, C2 – Inputs
O1, O2, O3 – Outputs

Truth Table:

A	B	C	O
0	0	0	0
0	0	1	1
0	1	0	1
0	1	1	1
1	0	0	1
1	0	1	1
1	1	0	1
1	1	1	1

Operation Mode:
- The gates are independent.
- The logic level found at the output of each gate depends on the logic levels applied to the inputs according to the truth table.

Electrical Characteristics:

Characteristic	Conditions (Vdd)	Value	Units
Drain/Source Current (typ)	5V	0.88	mA
	10V	2.2	mA
	15V	8	mA
Propagation Delay (typ)	5V	140	ns
	10V	70	ns
	15V	50	ns
Quiescent Device Current (max)	5V	0.25	µA
	10V	0.5	µA
	15V	1	µA
Supply Voltage Range	3 to 15		V

Other Devices:
- The 4071 is formed by 2-input OR gates, but is not buffered.

Applications:
- Logic Functions
- Digital Amplifiers
- Oscillators

Observations:
The outputs can drive two low-power TTL inputs or one TTL LS input.

4076

Tri-State Quad D Flip-Flop

Description: This device is formed by four D flip-flops with common clock, disable, clear input, and tri-state outputs.

Functional Diagram or/and Package:

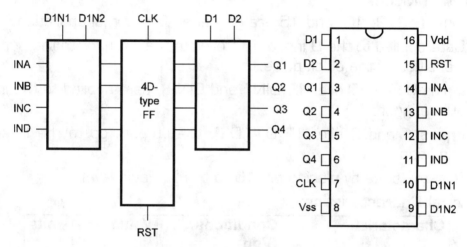

(Figure 2.59)

Pin Names:

Vdd – Positive Supply Voltage (3V to 15V)
Vss – Ground
INA, INB, INC, IND – Inputs
Q1, Q2, Q3, Q4 – Outputs
CLK – Clock
RST – Reset
D1, D2 – Disable output
DIN1, DIN2 – Disable input

Truth Table:

tn		tn+1
Data Input Disable	Data input	
Logic "1" on One or Both Inputs	X	Qn
Logic "0" on Both Inputs	1	1
Logic "0" on Both Inputs	0	0

Operation Mode:
- Pins 1, 2, 9, 10, and 15 are grounded for normal operation.
- Data applied to the D inputs is stored in the circuit with the positive transition of the clock pulse.
- If pins 9 and 10 are "1" (DIN1 and DIN2), data applied to the input will be ignored.
- If pins 1 and 2 are "1" (D1, D2), the outputs go to the tri-state condition.
- Reset is done by placing pin 15 to the "1" logic level.

Electrical Characteristics:

Characteristic	Conditions (Vdd)	Value	Units
Drain/Source Current (typ)	5V	0.88	mA
	10V	2.2	mA
	15V	8.8	mA
Maximum Clock Frequency (typ)	5V	4.0	MHz
	10V	12.0	MHz
	15V	15.0	MHz
Quiescent Device Current (max)	5V	5	µA
	10V	10	µA
	15V	20	µA
Supply Voltage Range	3 to 15		V

Applications:
- Interfacing
- Data Storage

4081

Quad 2-Input AND Buffered B Series Gate

Description: This device is formed by four 2-input buffered gates.

Functional Diagram and/or Package:

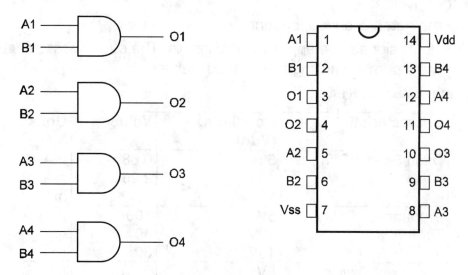

(Figure 2.60)

Pin Names:

Vdd – Positive Supply Voltage (3V to 15V)
Vss – Ground
A1, B1, A2, B2, A3, B3, A4, B4 – Inputs
O1, O2, O3, O4 – Outputs

Truth Table:

A	B	O
0	0	1
0	1	1
1	0	1
1	1	0

Operation Mode:
- All the gates are independent.
- The logic signals are applied to the inputs. The output state depends on the inputs according to the truth table.

Electrical Characteristics:

Characteristic	Conditions (Vdd)	Value	Units
Drain/Source Current	5V	0.88	mA
	10V	2.25	mA
	15V	8.8	mA
Propagation Delay Time	5V	100	ns
	10V	40	ns
	15V	30	ns
Quiescent Device Current (max)	5V	0.25	µA
	10V	0.5	µA
	15V	1	µA
Supply Voltage Range	3 to 15		V

Other Devices:
The 4001 is an equivalent device.

Applications:
- Interface and Driver
- Digital Amplifier
- High Voltage Applications (20V)

4082

Dual 4-Input AND Gates

Description: The two 4-input AND gates in this package can be used independently.

Functional Diagram and/or Package:

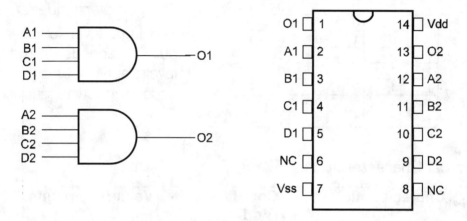

(Figure 2.61)

Pin Names:

Vdd – Positive Supply Voltage (3V to 15V)
Vss – Ground
A1, B1, C1, D1, A2, B2, C2 – Inputs
O1, O2 – Outputs
NC – Not Connected

Truth Table:

A	B	C	D	O
0	0	0	0	0
0	0	0	1	0
0	0	1	0	0
0	0	1	1	0
0	1	0	0	0
0	1	0	1	0
0	1	1	0	0
0	1	1	1	0
1	0	0	0	0
1	0	0	1	0
1	0	1	0	0
1	0	1	1	0
1	1	0	0	0
1	1	0	1	0
1	1	1	0	0
1	1	1	1	1

Operation Mode:
- The gates are independent.
- The logic level found in the output depends on the logic levels applied to the inputs according to the truth table.

Electrical Characteristics:

Characteristic	Conditions (Vdd)	Value	Units
Drain/Source Current (typ)	5V	0.88	mA
	10V	2.25	mA
	15V	8.8	mA
Propagation Delay Time (typ)	5V	125	ns
	10V	60	ns
	15V	45	ns
Quiescent Device Current (max)	5V	0.25	µA
	10V	0.5	µA
	15V	1	µA
Supply Voltage Range	3 to 15		V

Applications:
- Interfacing and Drive
- Digital Amplifier

4089

Binary Rate Multiplier

Description: This package contains a logic block that may be used to make the multiplication of a pulse rate by a selected amount.

Functional Diagram and/or Package:

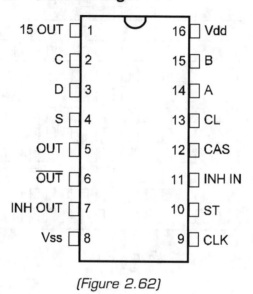

(Figure 2.62)

Pin Names:

Vdd – Positive Supply Voltage (3V to 15V)
Vss – Ground
CL – Clear
S – Set to 15
15-OUT – 15 output
CLK – Clock
OUT, OUT/ – Output and Complementary Output

INH OUT – Inhibit Output
INH IN – Inhibit Input
ST – Strobe
CAS – Cascade
A, B, C, D – Binary Input Number (D = MSB)

CMOS Sourcebook

Truth Table:

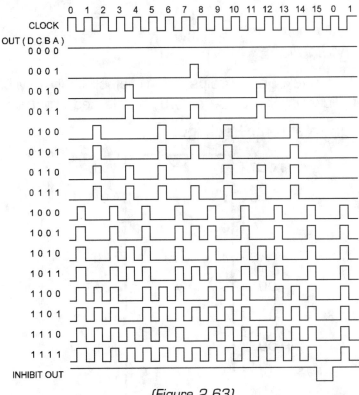

(Figure 2.63)

Operation Mode:

- Pins 4, 10, 11, 12, and 13 are grounded in normal operation.
- Clock is applied to pin 9.
- The binary number to which the multiplication will be made is applied to the inputs A, B, C, and D (D is the MSB).
- Output is obtained at pin 6. The complement is found in pin 5.
- The output is a signal with a frequency of one sixteenth of the input frequency multiplied by the word presented to pins A, B, C, and D.
- S and CL are used to synchronize the circuit.
- INH IN stops the output pulses when put to "1".

Functional Diagrams and Information for Designers

Electrical Characteristics:

Characteristic	Conditions (Vdd)	Value	Units
Drain/Source Current (typ)	5V	0.88	mA
	10V	2.25	mA
	15V	8.8	mA
Maximum Clock Frequency (typ)	5V	2	MHz
	10V	5	MHz
	15V	7	MHz
Quiescent Device Current (max)	5V	5	µA
	10V	10	µA
	15V	20	µA
Supply Voltage Range	3 to 15		V

Other Devices:
- The 4527 is equivalent except for decimal rates.

Applications:
- Instrumentation
- Numerical Control
- Digital Filters
- Frequency Synthesis

Observations:
- The output is average, which means that the pulses are not equally spaced.
- This device can be used in A/D and D/A converters.

4093

2-Input NAND Schmitt Trigger

Description: The 4-Input NAND Schmitt Triggers in this package can be used independently. They have as the main characteristic the hysteresis. The "snap" action (hysteresis) of the gates found in this device make it ideal for noisy or slow input voltage changes and as oscillator or monostable applications.

Functional Diagram and/or Package:

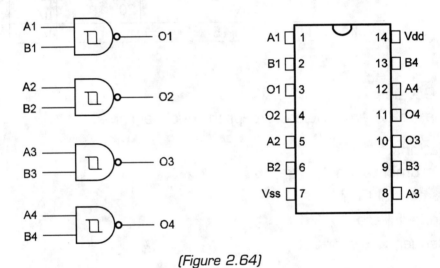

(Figure 2.64)

Pin Names:

Vdd – Positive Supply Voltage (3V to 15V)
Vss – Ground
A1, B1, A2, B2, A3, B3, A4, B4 – Inputs
O1, O2, O3, O4 – Outputs

Functional Diagrams and Information for Designers

Truth Table:

A	B	O
0	0	1
0	1	1
1	0	1
1	1	0

Operation Mode:
- The four gates are independent.
- The output logic level depends on the logic levels applied to the inputs according to the truth table.

Electrical Characteristics:

Characteristic	Conditions (Vdd)	Value	Units
Drain/Source Current (typ)	5V	0.88	mA
	10V	2.25	mA
	15V	8.8	mA
Propagation Delay Time (typ)	5V	300	ns
	10V	120	ns
	15V	80	ns
Quiescent Device Current (max)	5V	0.25	µA
	10V	0.5	µA
	15V	1	µA
Hysteresis (typ) - (see observations)	5V	1.6	V
	10V	2.2	V
	15V	2.7	V
Supply Voltage Range	3 to 15		V

Other Devices:
- If only the snap action of a Schmitt Trigger is needed, the equivalent Schmitt Inverter 40106 can be used.

Applications:
- Logic Functions (NAND and Inverter)
- Signal Conditioning
- Wave and Pulse Shapers
- Oscillators
- Digital Amplifiers
- Interfacing
- Monostable/Astable Multivibrators

Observations:
This device presents a "hysteresis characteristic." The hysteresis voltage or Vh is defined as the difference between the positive and the negative voltages when the device is triggered on and off.

4094

8-Bit Shift Register/Latch

Description: This package contains an 8-bit shift register and a tri-state 8-bit latch. The output of the last stage can be used to cascade several devices.

Functional Diagram and/or Package:

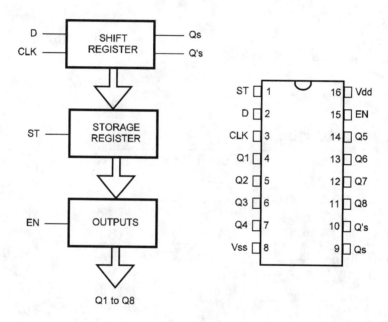

(Figure 2.65)

Pin Names:

Vdd – Positive Supply Voltage (3V to 15V)
Vss – Ground
Q1 to Q8 – Data Outputs
CLK – Clock

D – Data
ST – Strobe
EN – Output Enable
Q's – Complementary Output
Qs – Output

Truth Table:

CLK	EN	ST	D	Parallel Outputs		Serial Outputs	
				Q1	QN	Qs	Q's
↑	0	X	X	Hi-Z	Hi-Z	Q7	NC
↓	0	X	X	Hi-Z	Hi-Z	NC	Q7
↑	1	0	X	NC	NC	Q7	NC
↑	1	1	0	0	Qn-1	Q7	NC
↑	1	1	1	1	Qn-1	Q7	NC
↓	1	1	1	NC	NC	NC	Q7

↓↑ - Signal level transitions
NC – No change
X – Don't care
Hi-Z – High impedance state (tri-state)

Operation Mode:

- The parallel outputs are connected directly to common bus lines.
- Data is shifted with the positive transition of the clock signal.
- Data in each shift-register stage is transferred to the storage register when the strobe input ST is placed at "1".
- Data appears at the outputs whenever EN is "1".

Electrical Characteristics:

Characteristic	Conditions (Vdd)	Value	Units
Drain/Source Current (typ)	5V	0.88	mA
	10V	2.25	mA
	15V	8.8	mA
Maximum Clock Frequency (typ)	5V	3.0	MHz
	10V	6.0	MHz
	15V	8.0	MHz
Quiescent Device Current (max)	5V	5	µA
	10V	10	µA
	15V	20	µA
Supply Voltage Range	3 to 15		V

Applications:

- Serial to Parallel Data Conversion
- Remote Control Holding Register

Observations:

Two serial outputs are available by the use of cascaded 4094 devices.

Functional Diagrams and Information for Designers

4099

8-Bit Addressable Latch

Description: 8 bits can be addressed from three address inputs using the circuit inside this package.

Functional Diagram and/or Package:

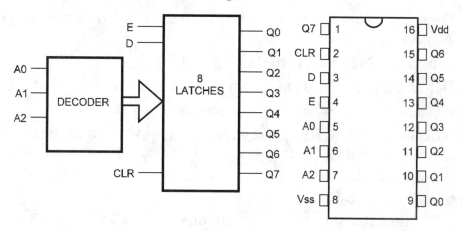

(Figure 2.66)

Pin Names:

Vdd – Positive Supply Voltage (3V to 15V)
Vss – Ground
Q0 to Q7 – Outputs
D – Data Input
CLR – Clear
A0, A1, A2 – Address Inputs
E – Enable

Truth Table:

E	CL	Addressed Latch	Unaddressed Latch	Mode
0	0	Follows Data	Holds Previous Data	Addressable Latch
1	0	Holds Previous Data	Holds Previous Data	Memory
0	1	Follows Data	Reset to 0	Demultiplexer
1	1	Reset to 0	Reset to 0	Clear

Operation Mode:

- Data enters by a particular bit in the latch when addressed by inputs A0, A1, and A2 and when E is "0".
- When E is "1", data entry is inhibited.
- Outputs can be read independently of the E state at any time.
- If CLR =1 and E =1 all the bits are reset to "0".
- If CLR =1 and E =0 the devices act as a 1-of-8 Demultiplexer. The bit addressed has an active output that follows the data input. The others are kept in "0".

Electrical Characteristics:

Characteristic	Conditions (Vdd)	Value	Units
Drain/Source Current (typ)	5V	0.88	mA
	10V	2.25	mA
	15V	8.8	mA
Propagation Delay (Data to Output) – (typ)	5V	200	ns
	10V	75	ns
	15V	60	ns
Quiescent Device Current (max)	5V	5	µA
	10V	10	µA
	15V	20	µA
Supply Voltage Range	3 to 15		V

Applications:

- Data Converters (Serial-to-Parallel)
- Remote Control Holding Register
- General Purpose Register

40106

Hex Schmitt Trigger

Description: This package contains six Schmitt Triggers (inverters). The snap action of this circuit makes it ideal for noisy or slow change signals and also for monostable and astable multivibrators.

Functional Diagram and/or Package:

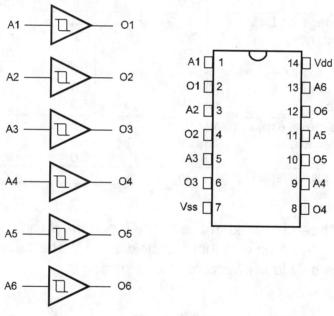

(Figure 2.67)

Pin Names:

Vdd – Positive Supply Voltage (3V to 15V)
Vss – Ground
A1, A2, A3, A4, A5, A6 – Inputs
O1, O2, O3, O4, O5, O6 – Outputs

CMOS Sourcebook

Truth Table:

A	O
0	1
1	0

Operation Mode:
- All inverters in this package are independent.
- The output of each inverter is "O" when the input is "1" and vice versa (see truth table).

Electrical Characteristics:

Characteristic	Conditions (Vdd)	Value	Units
Drain/Source Current (typ)	5V	0.88	mA
	10V	2.25	mA
	15V	8.8	mA
Propagation Delay Time (typ)	5V	220	ns
	10V	80	ns
	15V	70	ns
Quiescent Device Current (max)	5V	1.0	µA
	10V	2.0	µA
	15V	4.0	µA
Hysteresis (typ) – (see observations)	5V	2.2	V
	10V	3.6	V
	15V	5.0	V
Supply Voltage Range	3 to 15		V

Other Devices:
- The 4093 can be used in the same functions if the inputs are connected together or one input is held at "1". The difference is that you'll have only four inverters in one package.

Applications:
- Oscillators (Astable)
- Monostable Multivibrators
- Wave and Pulse Shapers
- Logic Functions

Observations:
This device presents a snap action when triggered. This is due the "hysteresis" or the difference between the positive voltage and the negative voltage.

40160

Decade Counter with Asynchronous Clear

Description: The synchronous counter found in this package features an internal carry look-ahead for fast counting and allows cascading packages without the need of additional logic.

Functional Diagram and/or package:

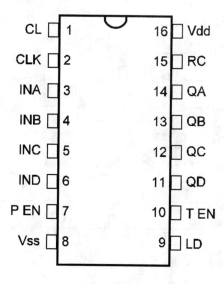

(Figure 2.68)

Pin Names:

Vdd – Positive Supply Voltage (3V to 15V)
Vss – Ground
CLK – Clock
CL – Clear
T EN, P EN – Enable
RC – Ripple Carry
LD – Load
INA, INB, INC, IND – Inputs
QA, QB, QC, QD – Outputs

Logic waveforms:

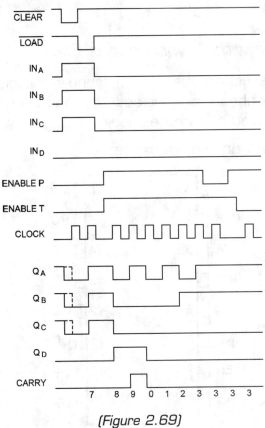

(Figure 2.69)

Operation Mode:
- The Clear (CL) function is asynchronous and when CL=1 the outputs are set to "0".
- When load is "0" (LD), the counter is disabled causing the output to agree with the setup data after the next clock pulse, regardless of the level of the Enable (T EN and P EN) inputs.
- Counting is enabled when T EN and P EN are "1".

Functional Diagrams and Information for Designers

Electrical Characteristics:

Characteristic	Conditions (Vdd)	Value	Units
Drain/Source Current (typ)	5V	0.88	mA
	10V	2.25	mA
	15V	8.8	mA
Maximum Clock Frequency (typ)	5V	4	MHz
	10V	11	MHz
	15V	14	MHz
Quiescent Device Current (max)	5V	5	µA
	10V	10	µA
	15V	20	µA
Supply Voltage Range	3 to 15		V

Other Devices:
- 40161, 40162, and 40163 are devices of the same group presenting only small differences in the operation mode.

Applications:
- Programmable Counters
- Frequency Dividers
- Timers

Observations:
This device is the equivalent CMOS of the TTL 74160.

40161

Binary Counter with Asynchronous Clear

Description: The synchronous counter in this package can be preset and also has an internal carry look-ahead for fast counting schemes and for cascading without the need of additional logic.

Functional Diagram and/or Package:

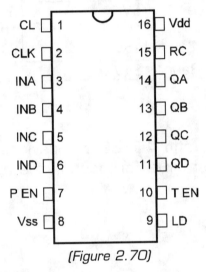

(Figure 2.70)

Pin Names:

    Vdd – Positive Supply Voltage (3V to 15V)
    Vss – Ground
    CLK – Clock
    CL – Clear
    P EN, T EN – Enable Inputs
    RC – Ripple Carry
    LD – Load
    QA, QB, QC, QD – Outputs
    INA, INB, INC, IND – Inputs

Functional Diagrams and Information for Designers

Logic waveforms:

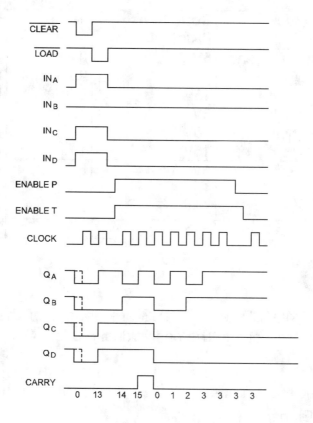

(Figure 2.71)

Operation Mode:

- The Clear (CL) function is asynchronous. CL=1 sets the outputs to "0".
- When Load (LD) is "0", the counter is disabled and causes the output to agree with the setup data after the next clock pulse, regardless of the levels in ENs inputs (P EN and T EN).
- Counting is enabled when P EN and T EN are "1".

Electrical Characteristics:

Characteristic	Conditions (Vdd)	Value	Units
Drain/Source Current (typ)	5V	0.88	mA
	10V	2.25	mA
	15V	8.8	mA
Maximum Clock Frequency (typ)	5V	4	MHz
	10V	11	MHz
	15V	14	MHz
Quiescent Device Current (max)	5V	5	µA
	10V	10	µA
	15V	20	µA
Supply Voltage Range	3 to 15		V

Other Devices:
- The 40160, 40162, and 40163 are devices of the same group with small differences in the operational mode.

Applications:
- Programmable Counters
- Frequency Dividers
- Timers

Observations:
This device is compatible with the TTL 74161.

40162

Decade Counter with Synchronous Clear

Description: The synchronous counter can be preset and also has an internal carry look-ahead for fast counting and to allow cascading packages without the need of additional logic.

Functional Diagram and/or Package:

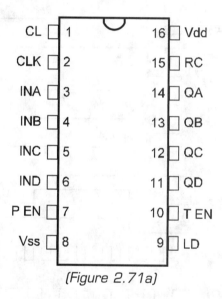

(Figure 2.71a)

Pin Names:

Vdd – Positive Supply Voltage (3V to 15V)
Vss – Ground
CLK – Clock
CL – Clear
P EN, T EN – Enable Inputs
LD – Load
INA, INB, INC, IND – Inputs
QA, QB, QC, QD – Outputs

Operation Mode:

- The Clear (CL) function is synchronous. CL=1 sets the outputs to "0".
- Load = 0 disables the counter and causes the output to agree with the setup data after the next clock pulse, regardless of the levels present in EM inputs (P and T).
- Counting is enabled when P EN and T EM = 1.

Logic Waveforms:

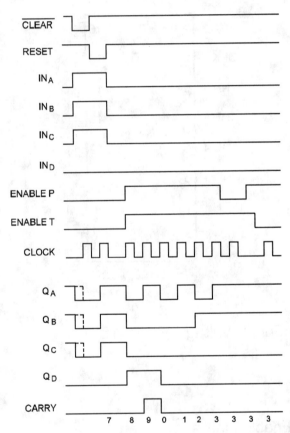

(Figure 2.72)

Electrical Characteristics:

Characteristic	Conditions (Vdd)	Value	Units
Drain/Source Current (typ)	5V	0.88	mA
	10V	2.25	mA
	15V	8.8	mA
Maximum Clock Frequency	5V	4	MHz
	10V	11	MHz
	15V	14	MHz
Quiescent Device Current (max)	5V	5	µA
	10V	10	µA
	15V	20	µA
Supply Voltage Range	3 to 15		V

Other Devices:
- The 40160, 40161, and 40163 are devices of the same group with only small differences in the operation mode.

Applications:
- Programmable Counters
- Frequency Dividers
- Timers

Observations:
This device is the CMOS equivalent to the TTL 74162.

40163

Binary Counter with Synchronous Clear

Description: The synchronous counter can be preset and also has an internal carry look-ahead for fast counting and to allow cascading without the need of additional logic.

Functional Diagram and/or Package:

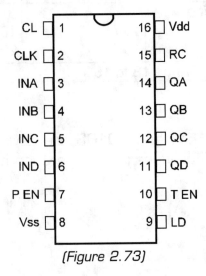

(Figure 2.73)

Pin Names:

Vdd – Positive Supply Voltage (3V to 15V)
Vss – Ground
CLK – Clock
CL – Clear
P EN, T EN – Enable
RC – Ripple Carry
INA, INB, INC, IND – Inputs
QA, QB, QC, QD – Outputs
LD – Load

Functional Diagrams and Information for Designers

Logic Waveforms:

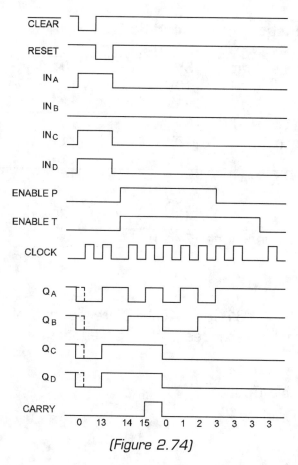

(Figure 2.74)

Operation Mode:

- The Clear (CL) function is synchronous. CL=1 sets the outputs to "0".
- LOAD (LD) = 0 disables the counter and causes the output to agree with the setup data after the next clock pulse, regardless of the levels present in the T EN and P EN inputs.
- Counting is enabled when P EN =T EN =1.

Electrical Characteristics:

Characteristic	Conditions (Vdd)	Value	Units
Drain/Source Current (typ)	5V	0.88	mA
	10V	2.25	mA
	15V	8.8	mA
Maximum Clock Frequency (typ)	5V	4	MHz
	10V	11	MHz
	15V	14	MHz
Quiescent Device Current (max)	5V	5	µA
	10V	10	µA
	15V	20	µA
Supply Voltage Range	3 to 15		V

Other Devices:
- The 40160, 40161, and 40162 are devices of the same group with only small differences in the operational mode.

Applications:
- Programmable Counters
- Frequency Dividers
- Timers

Observations:
This device is compatible with the equivalent TTL 74163.

Functional Diagrams and Information for Designers

40174

Hex D Flip-Flop

Description: Six D-type flip-flops with common CLEAR (CL) are found inside this package. The circuit can be used to store digital information.

Functional Diagram and Package:

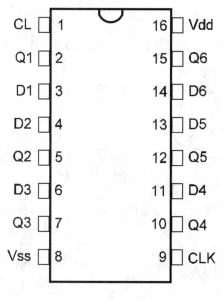

(Figure 2.75)

Pin Names:

    Vdd – Positive Supply Voltage (3 to 30V)
    Vss – Ground
    CL – Clear
    CLK – Clock
    Q1 to Q6 – Outputs
    D1 to D6 – D Inputs

CMOS Sourcebook

Truth Table:

CL	CLK	D	Q
0	X	X	0
1	↑	1	1
1	↑	0	0
1	1	X	NC
1	0	X	NC

↑ – Signal level transition
X – Don't care
NC – No Change

Operation Mode:

- Data is transferred to the Q outputs with the positive transition of the clock pulse.
- All flip-flops are reset to "0" when CL is put to the "0" logic level (see TruthTable).

Electrical Characteristics:

Characteristic	Conditions (Vdd)	Value	Units
Drain/Source Current (typ)	5V	0.88	mA
	10V	2.25	mA
	15V	8.8	mA
Maximum Clock Frequency (typ)	5V	3.5	MHz
	10V	10	MHz
	15V	12	MHz
Quiescent Device Current (max)	5V	1.0	µA
	10V	2.0	µA
	15V	4.0	µA
Supply Voltage Range	3 to 15		V

Applications:

- Shift Registers
- Pattern Generators
- Buffer/Storage Registers

Observations:

This device is equivalent to the TTL 74174.

40175

Quad D Flip-Flop

Description: Four D-type flip-flops with common CLEAR are found in this device. The circuit can be used to store digital information.

Functional Diagram and Package:

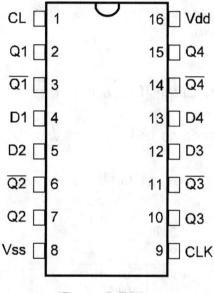

(Figure 2.76)

Pin Names:

Vdd – Positive Supply Voltage (3V to 15V)
Vss – Ground
CLK – Clock
CL – Clear
Q1 to Q4 – Normal Outputs
Q1/ to Q4/ – Complementary Outputs
D1 to D4 – D Inputs

Truth Table:

Inputs			Outputs	
CL	CLK	D	Q	Q/
0	X	X	0	1
1	↑	1	1	0
1	↑	1	0	1
1	1	X	NC	NC
1	0	X	NC	NC

↑ – Signal level transition
NC – No Change
X – Don't care

Operation Mode:

- Clock (CLK) and Clear (CL) are common to all flip-flops in this package.
- Data is transferred to the outputs with the positive transition of the clock.
- CL = 0 reset causes all flip-flops to reset (see Truth Table for more details).

Electrical Characteristics:

Characteristic	Conditions (Vdd)	Value	Units
Drain/Source Current (typ)	5V	0.88	mA
	10V	2.25	mA
	15V	8.8	mA
Maximum Clock Frequency (typ)	5V	3.5	MHz
	10V	10	MHz
	15V	12	MHz
Quiescent Device Current (max)	5V	1.0	µA
	10V	2.0	µA
	15V	4.0	µA
Supply Voltage Range	3 to 15		V

Functional Diagrams and Information for Designers

Applications:
- Shift Registers
- Pattern Generators
- Buffer and Storage Registers

Observations:
This device is the CMOS equivalent to the TTL 74192.

40192

4-Bit Up/Down Decade Counter

Description: The synchronous counter inside this package has separate clocks for up and down counting. The stages can be cascaded using the internal carry-borrow logic.

Functional Diagram and/or package:

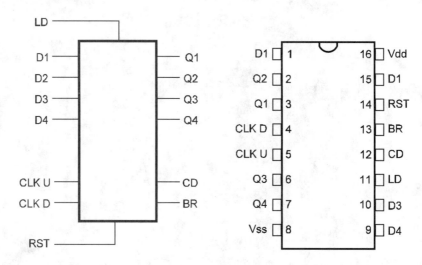

(Figure 2.77)

Pin Names:

Vdd – Positive Supply Voltage (3V to 15V)
Vss – Ground
CLK D, Clock U – Clock Down and Clock Up
CD – Cascade
RST – Reset
BR – Borrow
LD – Load
D1 to D4 – D Inputs
Q1 to Q4 – Outputs

Functional Diagrams and Information for Designers

Timing Diagram:

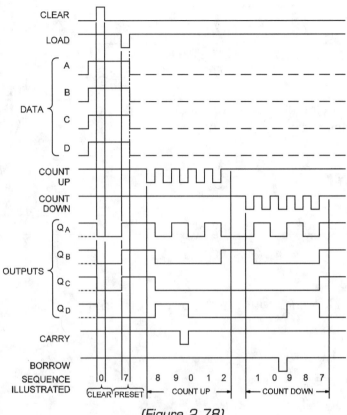

(Figure 2.78)

Operation Mode:

- For normal operation RS = 0 and LD = 1.
- Clock Down and Clock Up (LK D/CLK U) are held at "1".
- When Clock Up receives a negative pulse, the count advances one count. When the clock down receives a negative pulse, the count holds back one count. The count occurs in the positive edge of the pulse.
- Presetting to a desired count can be done using the D inputs and grounding by LD.
- Reset is done by setting RS to "1".

Electrical Characteristics:

Characteristic	Conditions (Vdd)	Value	Units
Drain/Source Current (typ)	5V	0.88	mA
	10V	2.25	mA
	15V	8.8	mA
Maximum Clock Frequency (typ)	5V	4	MHz
	10V	10	MHz
	15V	12.5	MHz
Quiescent Device Current (max)	5V	5	µA
	10V	10	µA
	15V	20	µA
Supply Voltage Range	3 to 15		V

Other Devices:
- The 40193 is the equivalent device for binary count.

Applications:
- Up/Down Counters
- Ripple Counters
- A/D and D/A Conversion
- Frequency Dividers
- Programmable BCD or Binary Counters

Observations:
Some data books also indicate this device as 4191.

40193

4-Bit Up/Down Binary Counter

Description: The synchronous counter inside this package has separate clocks for counting up and down. The stages may be cascaded using the internal carry-borrow logic.

Functional Diagram and/or Package:

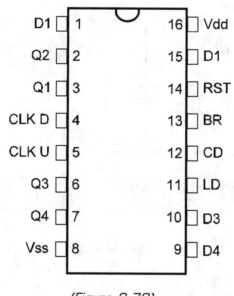

(Figure 2.79)

Pin Names:

Vdd – Positive Supply Voltage (3V to 15V)
Vss – Ground
CLK U, CLK D – Clock Up and Clock Down
CD – Cascade
LD – Load
RS – Reset
D1 to D4 – D Inputs
Q1 to Q4 – Outputs

CMOS Sourcebook

Timing Diagram:

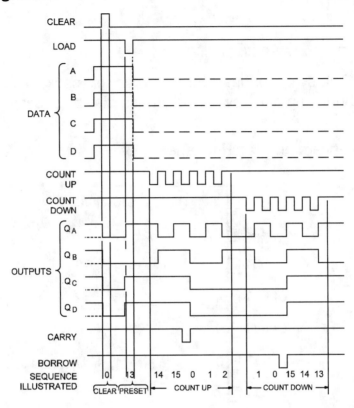

(Figure 2.80)

Operation Mode:

- Normal operation: RS = 0, LD = 1, CLK U = 1, CLK D = 1.
- When CLK U receives a negative pulse, the count advances one count with the positive edge of the pulse. When CLK D receives a negative pulse, the count holds back one count with the positive edge of the pulse.
- The counter can be preset to a predetermined count using the D inputs.
- When RS is put to "1", all the outputs go to "0". (0000)

Functional Diagrams and Information for Designers

Electrical Characteristics:

Characteristic	Conditions (Vdd)	Value	Units
Drain/Source Current (typ)	5V	0.88	mA
	10V	2.25	mA
	15V	8.8	mA
Maximum Clock Frequency (typ)	5V	4	MHz
	10V	10	MHz
	15V	12.5	MHz
Quiescent Device Current (max)	5V	5	µA
	10V	10	µA
	15V	20	µA
Supply Voltage Range	3 to 15		V

Other Devices:
- The 40192 is the equivalent decade counter.

Applications:
- Up/Down Counters
- Ripple Counters
- A/D and D/A Converters
- Frequency Dividers
- Programmable Binary or BCD Counters

Observations:
In some publications, this device is also called 4193.

4503

Non-Inverting Tri-State Buffer

Description: This device is formed by six non-inverting buffers with three-state outputs. The device has two separated disable inputs.

Functional Diagram and/or Package:

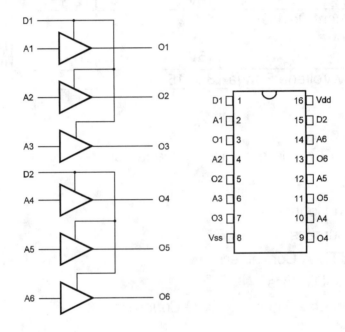

(Figure 2.81)

Pin Names:

Vdd – Positive Supply Voltage (3V to 15V)

Vss – Ground

A1 to A6 – Inputs

D1, D2 – Disable Inputs

O1 to O6 – Outputs

Functional Diagrams and Information for Designers

Truth Table:

In	Disable In	Output
0	0	0
1	0	1
X	1	Tri-state

X = Don't care

Operation Mode:
- The output of each gate depends on the inputs according to the truth table.
- When EN=1 the outputs go to the high-impedance state.

Electrical Characteristics:

Characteristic	Conditions (Vdd)	Value	Units
Drain/Source Current (typ)	5V	2.4/1.02	mA
	10V	8.3/2.60	mA
	15V	16.1/6.8	mA
Propagation Delay Time (typ)	5V	100	ns
	10V	40	ns
	15V	30	ns
Quiescent Device Current (max)	5V	1	µA
	10V	2	µA
	15V	4	µA
Supply Voltage Range	3 to 15		V

Other Devices:
- The 40502 is an equivalent device, but with a different pinout.

Applications:
- Interfacing
- Digital Amplifiers
- Logic Functions
- Logic Converters

Observations:
This device can drive two Standard TTL inputs or six LS TTL inputs.

4510

BCD Up/Down Counter

Description: This is a synchronous BCD up-down counter and divide-by-10 circuit with preset inputs. Packages may be cascaded using the internal carry/borrow logic.

Functional Diagram and/or Package:

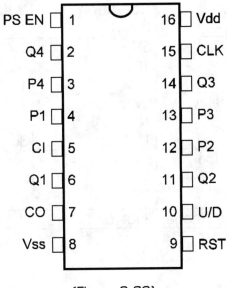

(Figure 2.82)

Pin Names:

Vdd – Positive Supply Voltage
Vss – Ground
CLK – Clock
P1, P2, P3, P4 – Parallel Inputs
U/D – Up-down Selection

PS EN – Preset Enable
CI – Carry In
CO – Carry Out
RST – Reset
Q1 to Q4 – Outputs

Functional Diagrams and Information for Designers

Truth Table:

CLK	RST	PSEN	CI	Up/Down	Output Function
X	1	X	X	X	Reset to zero
X	0	1	X	X	Set to P1, P2, P3, P4
↑	0	1	X	1	Count Up
↑	0	0	0	1	Count Down
↓	0	0	1	X	No Change

Operation Mode:

- For normal operation CI = 0, RST = 0 and PSEN = 0
- With U/D = 0 the count advances one count with the positive transition of the clock. With U/D = 1 the count holds back one count with the positive transition of the clock.
- Data may be parallel loaded using the A1 to A4 lines and applying a positive pulse to PSEN.
- Making RST = 1 the count will be reset to 0000.
- CLK must be "0" during loading and reset (see Truth Table for more details).

Electrical Characteristics:

Characteristic	Conditions (Vdd)	Value	Units
Drain/Source Current (typ)	5V	0.8	mA
	10V	2.0	mA
	15V	7.8	mA
Maximum Clock Frequency (typ)	5V	3.1	MHz
	10V	7.8	MHz
	15V	10.0	MHz
Quiescent Device Current (max)	5V	5	µA
	10V	10	µA
	15V	20	µA
Supply Voltage Range	3 to 15		V

Other Devices:
- The equivalent device for binary counting is the 4516.

Applications:
- Difference Counters
- Synchronous Multistage Counters
- Ripple Counters
- Frequency Dividers

Observations:
The Carry In (CI) input can also be used as an enable input. When high, the count is inhibited.

4516

Binary Up/Down Counter

Description: This package contains a binary up/down counter.

Functional Diagram and/or Package:

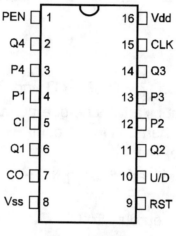

(Figure 2.83)

Pin Names:

    Vdd – Positive Supply Voltage (3V to 15V)
    Vss – Ground
    CLK – Clock
    RST – Reset
    PEN – Preset Enable
    CI – Carry In
    U/D – Up/Down
    P1, P2, P3, P4 – Parallel Inputs
    Q1, Q2, Q3, Q4 – Outputs
    CO – Carry Out

Truth Table:

CLK	RS	PEN	CI	U/D	Output Function
X	1	X	X	X	Reset to Zero
X	0	1	X	X	Set to P1, P2, P3, P4
↑	0	0	0	1	Count Up
↑	0	0	0	0	Count Down
↓	0	0	X	X	No Change
X	0	0	1	X	No Change

Operation Mode:

- For normal operation CI = 0, RS = 0 and PEN = 0.
- When U/D = 0, count advances one count with the positive transition of the clock. When U/D = 1, count retards one count with the positive transition of the clock.
- Data may be parallel loaded using the P1 to P4 inputs and applying a positive pulse to the PEN input.
- RST high brings the outputs to 0000.

Electrical Characteristics:

Characteristic	Conditions (Vdd)	Value	Units
Drain/Source Current (typ)	5V	0.8	mA
	10V	2.0	mA
	15V	7.8	mA
Maximum Clock Frequency	5V	3.1	MHz
	10V	7.6	MHz
	15V	10.0	MHz
Quiescent Device Current (max)	5V	5	µA
	10V	10	µA
	15V	20	µA
Supply Voltage Range	3 to 15		V

Other Devices:
- The 4510 is the equivalent decade counter.

Applications:
- Difference Counters
- Frequency Dividers
- Synchronous Counters
- Ripple Counters

Observations:
- The device can be reset loading 0000.
- Enable can be made using the CI input. CI = 1 inhibits the count.

4511

BCD-to-7 Segment Latch/Decoder/Driver

Description: This package converts BCD input codes by sorting them and converts them to a 7-segment high-current drive signal. The positive logic makes it suitable with common cathode 7-segment LED displays.

Functional Diagram and/or Package:

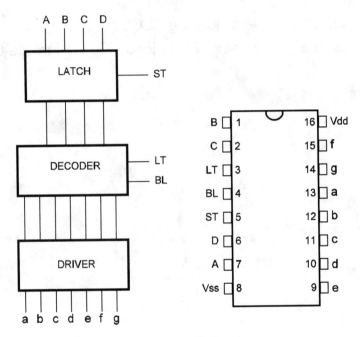

(Figure 2.84)

Pin Names:

Vdd – Positive Supply Voltage (3V to 15V)
Vss – Ground
a, b, c, d, e, f, g – Segment Outputs

1, 2, 4 – BCD Inputs
ST – Store
LT – Lamp Test
BL – Blanking

Functional Diagrams and Information for Designers

Truth Table:

Inputs							Outputs							Display
LE	BI	LT	D	C	B	A	a	b	c	d	e	f	g	
X	X	0	X	X	X	X	1	1	1	1	1	1	1	B
X	0	1	X	X	X	X	0	0	0	0	0	0	0	-
0	1	1	0	0	0	0	1	1	1	1	1	1	0	0
0	1	1	0	0	0	1	0	1	1	0	0	0	0	1
0	1	1	0	0	1	0	1	1	0	1	1	0	1	2
0	1	1	0	0	1	1	1	1	1	1	0	0	1	3
0	1	1	0	1	0	0	0	1	1	0	0	1	1	4
0	1	1	0	1	0	1	1	0	1	0	0	1	1	5
0	1	1	0	1	1	0	0	0	1	1	1	1	1	6
0	1	1	0	1	1	1	1	1	1	0	0	0	0	7
0	1	1	1	0	0	0	1	1	1	1	1	1	1	8
0	1	1	1	0	0	1	1	1	1	0	1	1	1	9
0	1	1	1	0	1	0	0	0	0	0	0	0	0	-
0	1	1	1	0	1	1	0	0	0	0	0	0	0	-
0	1	1	1	1	0	0	0	0	0	0	0	0	0	-
0	1	1	1	1	0	1	0	0	0	0	0	0	0	-
0	1	1	1	1	1	0	0	0	0	0	0	0	0	-
0	1	1	1	1	1	1	0	0	0	0	0	0	0	-
1	1	1	X	X	X	X	(*)							(*)

X – Don't care
(*) Depends on the BCD code supplied during the 0 to 1 transition of LE

Operation Mode:

- Normal operation: LT = 1, BL = 1 and ST = 0.
- The BCD code applied to the BCD Inputs (1, 2, and 4) results in the correspondent 7-segment levels in the output according to the truth table.
- The current in each segment must be limited by a 150-ohm resistor or higher.
- The outputs source current driving common cathode displays.

- Making ST = 1 the device retains the last BDC value applied to the input.
- Making BL = 0 all the outputs go to "0".
- Making LT = 0 all the segments will be activated for test.

Electrical Characteristics:

Characteristic	Conditions (Vdd)	Value	Units
Drain/Source Current (typ)	5V	0.88	mA
	10V	2.25	mA
	15V	8.8	mA
Output Rise Time (typ)	5V	80	ns
	10V	60	ns
	15V	50	ns
Quiescent Device Current (max)	5V	5	µA
	10V	10	µA
	15V	20	µA
Supply Voltage Range	3 to 15		V

Applications:
- Common Cathode LED Display Driver
- Multiplexing Displays
- Low Voltage Fluorescent Display Driver
- Incandescent Display Driver

Observations:
- BL can function as a brightness control by using a variable duty-cycle oscillator. The duty cycle of the applied signal will determine the brightness of the segment.
- Shorts in the outputs can damage the device.

4512

8-Channel Buffered Data Selector

Description: One of eight inputs can be selected and appear in the output of the circuit in this package. The device can be used as a data selector or to generate any four-variable logic function. The circuit has a tri-state output

Functional Diagram and Package:

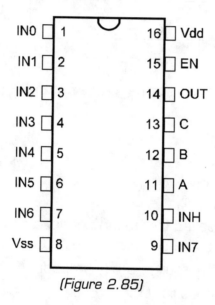

(Figure 2.85)

Pin Names:

Vdd – Positive Supply Voltage (3V to 15V)
Vss – Ground
IN0 to IN7 – Inputs
INH – Inhibit
A, B, C – Input Selector
OUT – Output EN – Enable

Truth Table:

Address Inputs			Control Inputs		Output
C	B	A	Inhibit	EN	OUT
0	0	0	0	0	IN0
0	0	1	0	0	IN1
0	1	0	0	0	IN2
0	1	1	0	0	IN3
1	0	0	0	0	IN4
1	0	1	0	0	IN5
1	1	0	0	0	IN6
1	1	1	0	0	IN7
X	X	X	1	0	0
X	X	X	X	1	Hi-Z

X – Don't care
Hi-Z – high impedance state (Tri-state condition)

Operation Mode:
- Normal operation: INH = 0.
- Select code is applied to the input selector (A to C where C = MSB). The bit applied to the selected input appears in the output.
- EN = 1 puts the output in a high-impedance state.
- INH = 1 puts output to "0".

Electrical Characteristics:

Characteristic	Conditions (Vdd)	Value	Units
Drain/Source Current (typ)	5V	0.51/0.	mA
	10V	1.3/0.5	mA
	15V	3.4/1.5	mA
Propagation Delay Time (typ)	5V	225	ns
	10V	75	ns
	15V	57	ns
Quiescent Device Current (max)	5V	5	µA
	10V	10	µA
	15V	20	µA
Supply Voltage Range	3 to 15		V

Applications:
- Digital Multiplexer
- Signal Gating
- Number-Sequence Generator

Observations:
This device is only a selector and can't be reversed to function as a data distributor.

4514/4515

4-Bit Latched/4-to-16 Line Decoder

Description: These packages contain a latch and a 1-of-16 decoder. The circuit may be used to select one output as a decoder or also to make the distribution of data to one of 16 outputs. The latches inside these packages are R-S flip-flops. The difference between the 4514 and the 4515 is that one is for logic "1" output and the other for logic "0" (see Truth Table).

Functional Diagram and/or Package:

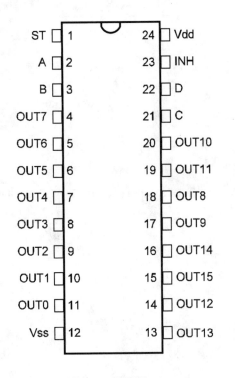

Pin Names:

Vdd – Positive Supply Voltage
Vss – Ground
OUT1 to OUT16 – Outputs
INH – Inhibit
A, B, C, D – Output Selector
ST – Strobe

(Figure 2.86)

Functional Diagrams and Information for Designers

Truth Table:

INH	Data Inputs				Selected Output 4514 – Logic 1 4515 – Logic 0
	A	B	C	D	
0	0	0	0	0	S0
0	0	0	0	1	S1
0	0	0	1	0	S2
0	0	0	1	1	S3
0	0	1	0	0	S4
0	0	1	0	1	S5
0	0	1	1	0	S6
0	0	1	1	1	S7
0	1	0	0	0	S8
0	1	0	0	1	S9
0	1	0	1	0	S10
0	1	0	1	1	S11
0	1	1	0	0	S12
0	1	1	0	1	S13
0	1	1	1	0	S14
0	1	1	1	1	S15
1	X	X	X	X	All outputs 0 (4514) All outputs 1 (4515)

X = Don't care

Operation Mode:

- Normal operation: ST=1 and INH = 0.
- The code applied to the output selector (A to D) selects the output. The selected output goes to "1" and the other remains in "0" for the 4514, or the selected output goes to "0" and the other remain in "1" for the 4515.
- INH = 1 makes all output go to "0" in the 4514 and to "1" in the 4515.
- ST = 0 stores the states of the input lines (see Truth Table for more details).

Electrical Characteristics:

Characteristic	Conditions (Vdd)	Value (typ)	Units
Drain/Source Current	5V	0.88	mA
	10V	2.25	mA
	15V	8.8	mA
Transition Time (typ)	5V	100	ns
	10V	50	ns
	15V	40	ns
Quiescent Device Current (max)	5V	5	µA
	10V	10	µA
	15V	20	µA
Supply Voltage Range	3 to 15		V

Applications:
- Digital Multiplexers
- Control Decoders
- Hexadecimal/BCD Decoders
- Address Decoders

Observations:
Changes in the select code will appear immediately in the outputs when F = 1.

4518

Dual Synchronous Up Counters

Description: This device is formed by two independent BCD synchronous up counters.

Functional Diagram and/or Package:

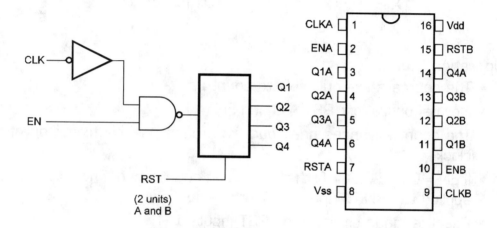

(Figure 2.86a)

Pin Names:

Vdd – Positive Supply Voltage (3V to 15V)
Vss – Ground
RST – Reset
CLK – Clock
EN – Enable
Q1, Q2, Q3, Q4 – Outputs

Truth Table:

CLK	EN	RS	Action
↑	1	0	Increment Counter
0	↓	0	Increment Counter
↓	X	0	No Change
X	↑	0	No Change
↑	0	0	No Change
1	↓	0	No Change
X	X	1	Q1 to Q4 = 0

X = Don't care

Operation Mode:
- The counters are only able to count up.
- Normal operation: RST = 0 and EN = 1.
- The count advances one count with each positive transition of the clock.
- If EN = 0 and CLK = 0 the circuit can count when a negative pulse is applied to the EN input (seeTruth Table).
- Reset is made bringing the RST input to "1".

Electrical Characteristics:

Characteristic	Conditions (Vdd)	Value	Units
Drain/Source Current (typ)	5V	0.88	mA
	10V	2.25	mA
	15V	8.8	mA
Maximum Clock Frequency (typ)	5V	3	MHz
	10V	6	MHz
	15V	8	MHz
Quiescent Device Current (max)	5V	5	µA
	10V	10	µA
	15V	20	µA
Supply Voltage Range	3 to 15		V

Functional Diagrams and Information for Designers

Other Devices:
- The 4520 is a binary counter with the same characteristics.

Applications:
- Multistage Counters
- Frequency Dividers
- Ripple Counters

Observations:
- To cascade the units, the 9 state (9 and 1) is applied to the EN input of the next counter.
- The output is a 1-2-4-8 binary code.

4520

Dual Synchronous Divide-by-16 Counter

Description: This device is formed by two independent binary up counters. These counters cannot be preset.

Functional Diagram and/or Package:

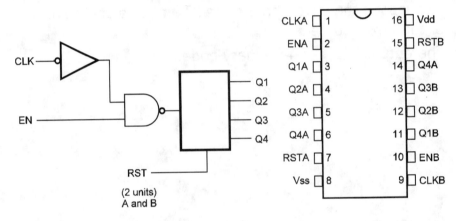

(Figure 2.87)

Pin Names:

Vdd – Positive Supply Voltage (3V to 15V)
Vss – Ground
Q1, Q2, Q3, Q4 – Outputs
CL – Clock
EN – Enable
RST – Reset

Functional Diagrams and Information for Designers

Truth Table:

CLK	EN	RST	Action
↑	1	0	Increment Counter
0	↓	0	Increment Counter
↓	X	0	No Change
X	↑	0	No Change
↑	0	0	No Change
1	↓	0	No Change
X	X	1	Q1 thru Q4 = 0

X = Don't care   ↓↑ = Signal transitions

Operation Mode:

- Normal Operation: RST = 0 and EN = 1.
- The counter advances one count with the positive transition of the clock.
- If RST = 0 and CL = 0 the counter will advance one count with a negative transition in the EN input.
- RST = 1 puts all the outputs to "0".

Electrical Characteristics:

Characteristic	Conditions (Vdd)	Value	Units
Drain/Source Current (typ)	5V	0.88	mA
	10V	2.25	mA
	15V	8.8	mA
Propagation Delay Time (typ)	5V	180	MHz
	10V	75	MHz
	15V	60	MHz
Quiescent Device Current (max)	5V	1	µA
	10V	2	µA
	15V	4	µA
Supply Voltage Range	3 to 15		V

Other Devices:
- The 4518 is a decade counter with the same basic characteristics.

Applications:
- Logic Functions
- Logic Control
- Multiplexers

Observations:
Cascading is done by detecting the 15 count with an AND gate (1 and 2 and 4 and 8) and applying the output to the EN of the next stage.

4522/4526

Programmable Divide-by-N 4-Bit Binary/Decimal Counters

Description: These devices are programmable and can be cascaded down counters with a decoded "0" state output for divide-by-n applications. The 4522 is a decimal counter and the 4526 a binary counter. The differences in the counting mode can be seen in the truth tables of the devices.

Functional Diagram or/and Package:

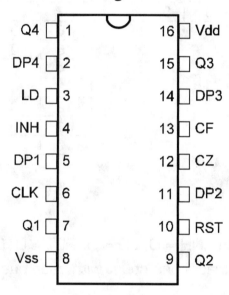

(Figure 2.88)

Pin Names:

Vdd – Positive Supply Voltage (3V to 15V)
Vss – Ground
CF – Cascade Feedback
RST – Master Reset
INH – Inhibit
CLK – Clock
DP1 to DP4 – BCD Code Input for Programming
Q1 to Q4 – Outputs
LD – CZ Load – Count Zero

CMOS Sourcebook

Truth Table:

a) 4522

Count	Q4	Q3	Q2	Q1
9	1	0	0	1
8	1	0	0	0
7	0	1	1	1
5	0	1	0	1
4	0	1	0	0
3	0	0	1	1
2	0	0	1	0
1	0	0	0	1
0	0	0	0	0

b) 4526

Count	Q4	Q3	Q2	Q1
15	1	1	1	1
14	1	1	1	0
13	1	1	0	1
12	1	1	0	0
11	1	0	1	1
10	1	0	1	0
9	1	0	0	1
8	1	0	0	0
7	0	1	1	1
6	0	1	1	0
5	0	1	0	1
4	0	1	0	0
3	0	0	1	1
2	0	0	1	0
1	0	0	0	1
0	0	0	0	0

Operation Mode:

- Normal operation: INH = 0, LD = 0, RST = 0, CF = 1
- The counter subtracts one count with each positive transition of the clock.
- A count can be loaded using the DP1, DP2, DP3, and DP4 inputs and making LD = 1. The code applied to these inputs is BCD (4522) or binary (4526).
- If CZ is connected to LD, the counter will proceed with the division by the count stored in the device even when the count reaches zero.
- RST = 1 will bring all the outputs to "0".

Functional Diagrams and Information for Designers

Electrical Characteristics:

Characteristic	Conditions (Vdd)	Value	Units
Drain/Source Current (typ)	5V	0.88	mA
	10V	2.25	mA
	15V	8.8	mA
Maximum Clock Frequency (typ)	5V	2.9	MHz
	10V	7.7	MHz
	15V	11	MHz
Quiescent Device Current (max)	5V	5	µA
	10V	10	µA
	15V	20	µA
Supply Voltage Range	3 to 15		V

Applications:
- Programmable Counters
- Frequency Dividers
- Timers

4528

Dual Monostable Multivibrator

Description: This package contains two independent monostable multivibrators, each with its own trigger and reset inputs.

Functional Diagram and/or Package:

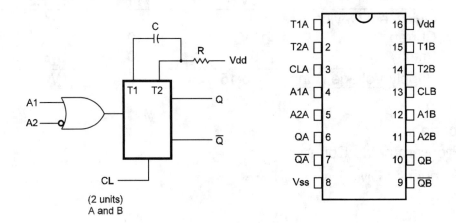

(Figure 2.89)

Pin Names:

Vdd – Positive Supply Voltage (3V to 15V)
Vss – Ground
A1, A2 – Inputs
T1A, T2A, T1B, T2B – Timing Capacitor and Resistor
Q1, Q2 – Outputs
Q1/, Q2/ – Complementary Outputs
C1, C2 – Timing Capacitor Inputs
CL – Clear

Functional Diagrams and Information for Designers

Truth Table:

Inputs			Outputs	
Clear	A	B	Q	Q/
0	X	X	0	1
X	1	X	0	1
X	X	0	0	1
1	0	↓	⇑	⇓
1	↑	1	⇑	⇓

↑ = Transition from 0 to 1
↓ = Transition from 1 to 0
⇑ = One high level pulse
⇓ = One low level pulse
X – Don't care

Operation Mode:

- Normal Operation: CL = 1 and the RC network connected as shown in Figure 2.89.
- Triggering:
    a) With positive pulses, use +IN and make -IN = 1.
    b) With negative pulses, use -IN and make +IN = 0.
- The on time is cut if CL is put to "0".

Electrical Characteristics:

Characteristic	Conditions (Vdd)	Value	Units
Drain/Source Current	5V	0.88/0.36	mA
	10V	2.25/0.9	mA
	15V	8.8/3.5	mA
Output Rise Time (typ)	5V	180	ns
	10V	90	ns
	15V	65	ns
Quiescent Device Current (max)	5V	5	µA
	10V	10	µA
	15V	20	µA
Supply Voltage Range	3 to 15		V

Other Devices:

- Use the 7555 (CMOS) or 555 (Bipolar) for long time delays.

Applications:
- Signal Conditioning
- Timers
- Missing Pulse Detectors
- Monostable Multivibrators

Observations:
- R can assume values between 10 k ohms and 10 M ohms.
- C must be higher than 20 pF.
- T (in seconds) is given by the next formula:

    $T = R \times C$

    C in Farads and R in ohms

4529

Dual 4-Channel or Single 8-Channel Analog Data Selector

Description: This device contains two analog data selectors. The device can be selected to operate as two 4-channel data selectors or one 8-channel data selector. The operation mode is selected by a code applied to an appropriated input. The device is suitable for digital as well as analog applications.

Functional Diagram and/or Package:

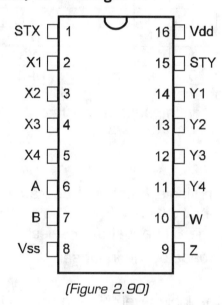

(Figure 2.90)

Pin Names:

Vdd – Positive Supply Voltage (3V to 15V)
Vss – Ground
Z, W – Outputs
X1 to X4, Y1 to Y4 – Inputs
A, B – Input Selection
STX, STY – State Inputs

Truth Table:

STX	STY	B	A	Z	W
1	1	0	0	X0	Y0
1	1	0	1	X1	Y1
1	1	1	0	X2	Y2
1	1	1	1	X3	Y3
1	0	0	0	X0	
1	0	0	1	X1	
1	0	1	0	X2	
1	0	1	1	X3	
0	1	0	0	Y0	
0	1	0	1	Y1	
0	1	1	0	Y2	
0	1	1	1	Y3	
0	0	X	X	Hi-Z	

X – Don't care
Hi-Z – High Impedance (tri-state)

Operation Mode:

- Dual 4-channel or single 8-channel operation mode is selected by the Z and W inputs. When W and Z are tied together, the device operates as a single 8-channel data selector.
- Input selection is made by the input selection inputs according to the truth table.
- STX and STY in the logic level "0" bring the outputs of the device to the high impedance state.

Functional Diagrams and Information for Designers

Electrical Characteristics:

Characteristic	Conditions (Vdd)	Value	Units
Drain/Source Current (typ)	5V	0.4	mA
	10V	1.1	mA
	15V	2.2	mA
On Resistance (typ)	5V	150	Ω
	10V	100	Ω
	15V	75	Ω
Quiescent Device Current (max)	5V	1.0	µA
	10V	1.0	µA
	15V	2.0	µA
Supply Voltage Range	3 to 15		V

Other Devices:
- The 4052 is a dual 4-channel Mux/Demux. The 4512 is an 8-channel data selector.

Applications:
- Multiplexing/Demultiplexing
- Logic
- Interfacing

Observations:
- Data paths are bidirectional.
- It can be used as a 1-of-4 or 1-of-8 decoder.

4538

Dual Precision Monostable

Description: This package contains a dual precision monostable multivibrator with independent trigger and reset controls. The device can be retriggered and reset, and the inputs are internally latched. The devices are also provided with two trigger inputs to trigger with rising and falling edge signals.

Functional Diagram and/or Package:

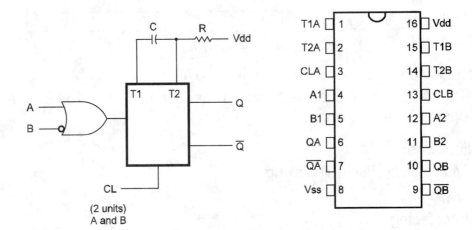

(Figure 2.91)

Pin Names:

Vdd – Positive Supply Voltage (3V to 15V)

Vss – Ground

CL1, CL2 – Clear

A1, B1, A2, B2 – Inputs

Q1, Q1/, Q2, Q2/ – Outputs

T1A, T2A, T1B, T2B – Timing Capacitors

Functional Diagrams and Information for Designers

Truth Table:

Inputs			Outputs	
CL	A	B	Q	Q/
0	X	X	0	1
X	1	X	0	1
X	X	0	0	1
1	0	↓	⇑	⇓
1	↑	1	⇑	⇓

⇑⇓ = One pulse
↑↓ = Transition level
X = Don't care

Operation Mode:

- When A is used to trigger the circuit with the positive edge of the input, signal B must be "1".
- If B is used to trigger the circuit with the negative edge of the pulses, A is grounded.
- C and R are connected, as shown in Figure 2.91.
- In a non-retriggerable version, Q/ is connected to B if the pulses enter by A and Q is connected to A if the pulses enter by B.

Electrical Characteristics:

Characteristic	Conditions (Vdd)	Value	Units
Drain/Source Current (typ)	5V	0.88	mA
	10V	2.25	mA
	15V	8.8	mA
Output Transition Time (typ)	5V	100	ns
	10V	50	ns
	15V	40	ns
Quiescent Device Current (max)	5V	5	µA
	10V	10	µA
	15V	20	µA
Supply Voltage Range	3 to 15		V

Other Devices:
- Use the 555 for long time delays (555 bipolar or 7555 CMOS).

Applications:
- Timers
- Monostable Multivibrators
- Signal Conditioning
- Missing Pulse Detectors

Observations:
An unused section must have A = 0, B = 1, CD = 1, T1= 0 and the other terminals unconnected.

Functional Diagrams and Information for Designers

4541

Programmable Timer

Description: This device is formed by a programmable timer designed with a 16-stage binary counter. The device is also equipped with an integrated oscillator for use with an external capacitor and two resistors, output control logic, and a special power-on reset circuit.

Functional Diagram and/or Package:

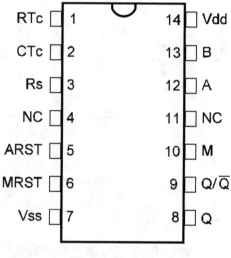

(Figure 2.92)

Pin Names:

Vdd – Positive Supply Voltage (3V to 15V)
Vss – Ground
A, B – Trigger Inputs
RTc, CTc, Rs – Timer Component Inputs

ARST – Auto Reset
MRST – Master Reset
M – Mode
Q – Output
Q/$\overline{Q}$ – Output Select

Truth Tables:

a) Truth Table

Pin	State	
	0	1
5	Auto Reset Operating	Auto Reset Disabled
6	Timer Operational	Master Reset On
9	Output Initially Low after Reset	Output Initially High after Reset
10	Single Cycle Mode	Recycle Mode

b) Division Rate Table

A	B	Number of Counter Stages	Count 2^n
0	0	13	8192
0	1	10	1024
1	0	8	256
1	1	16	65536

Operation Mode:

- Timing and count are initialized when the power is turned on.
- The Master Reset can be used to bring the circuit to "0" count. See Truth Table for more details.
- The frequency is determined by the external RC network.
- Division rate is determined by the inputs A and B (see Truth Table).

Functional Diagrams and Information for Designers

Electrical Characteristics:

Characteristic	Conditions (Vdd)	Value	Units
Drain/Source Current (typ)	5V	3.6/13	mA
	10V	9.0/8.0	mA
	15V	34/30	mA
Maximum Clock Frequency (typ)	5V	2.5	MHz
	10V	6.0	MHz
	15V	8.5	MHz
Quiescent Device Current (max)	5V	5	µA
	10V	10	µA
	15V	20	µA
Supply Voltage Range	3 to 15		V

Other Devices:
- The 555 and a frequency divider such as the 4020 or 4040 are options for a large time delay application replacing this device.

Applications:
- Long Period Timers
- Frequency Dividers
- Driver for 6 TTL LS

Observations:
It is recommended that C>20 pF and 10k<R<10 M ohms.

4543

BCD-to-7 Segment Latch/Decoder/Driver for Liquid Crystals

Description: This package contains a BCD-to-7-Segment Latch/Decoder/Driver for LCDs and other types of displays. The circuit provides the functions of a 4-bit storage latch and an 8421 BCD-to-7-segment decoder and driver. The device is able to invert the logic levels of the output combination.

Functional Diagram and Package:

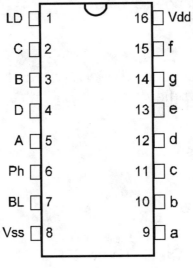

(Figure 2.93)

Pin Names:

Vdd – Positive Supply Voltage (3V to 15V)
Vss – Ground
a, b, c, d, e, f, g – Segment Outputs

A, B, C, D – BCD Inputs
LD – Latch Disable
Ph – Square Wave Input
BL – Blanking

Functional Diagrams and Information for Designers

Truth Table:

Inputs				Outputs	
LD	BL	Ph	D C B A	A b c d e f g	Display
1	0	0	X X X X	0 0 0 0 0 0 0	Blank
1	0	0	0 0 0 0	1 1 1 1 1 1 0	0
1	0	0	0 0 0 1	0 1 1 0 0 0 0	1
1	0	0	0 0 1 0	1 1 0 1 1 0 1	2
1	0	0	0 0 1 1	1 1 1 1 0 0 1	3
1	0	0	0 1 0 0	0 1 1 0 0 1 1	4
1	0	0	0 1 0 1	1 0 1 1 0 1 1	5
1	0	0	0 1 1 0	1 0 1 1 1 1 1	6
1	0	0	0 1 1 1	1 1 1 0 0 0 0	7
1	0	0	1 0 0 0	1 1 1 1 1 1 1	8
1	0	0	1 0 0 1	1 1 1 1 0 1 1	9
1	0	0	1 0 1 0	0 0 0 0 0 0 0	Blank
1	0	0	1 0 1 1	0 0 0 0 0 0 0	Blank
1	0	0	1 1 0 0	0 0 0 0 0 0 0	Blank
1	0	0	1 1 0 1	0 0 0 0 0 0 0	Blank
1	0	0	1 1 1 0	0 0 0 0 0 0 0	Blank
1	0	0	1 1 1 1	0 0 0 0 0 0 0	Blank
0	0	0	X X X X	(*)	(*)
(1)	(1)	1	(1)	Inverse of Output Combination Above	Display as Above

X – Don't care
(*) – Depends on the BCD code previously applied when LD=1
(1) – Above combinations

Operation Mode:
- Normal Operation: LD = 0, BL = 0 and Ph = 0.
- The BCD data applied to the BCD inputs results in logic levels in the segment outputs according to the truth table, generating the numbers in the display.
- LD = 0 makes the display to go to "0" in all segment outputs.
- Ph is used for the low-frequency signal necessary to drive LCDs (Liquid Crystal Display).
- BL = 1 put zeros in all segment outputs (see Truth Table).
- The Ph, BL, and LD can be used to reverse the truth table phase, blank the display, and store a BCD code, respectively.

Electrical Characteristics:

Characteristic	Conditions (Vdd)	Value	Units
Drain/Source Current (typ)	5V	0.51	mA
	10V	1.3	mA
	15V	3.4	mA
Output Rise Time (typ)	5V	100	ns
	10V	50	ns
	15V	40	ns
Quiescent Device Current (max)	5V	5	µA
	10V	10	µA
	15V	20	µA
Supply Voltage Range	3 to 15		V

Applications:
- LCD Driver
- Timing Device Driver (Clocks, Watches, Timers, etc.)

Observations:
This device can also be used to drive LED displays, incandescent, and other types using the appropriate interface.

4723

Dual 4-Bit Addressable Latch

Description: This device is formed by two 4-bit addressable latches with common control inputs. Other features include two address inputs (A0, A1) and an active high clear input (CL). Each latch has a data input (D) and four outputs (Q0 to Q3).

Functional Diagram and/or Package:

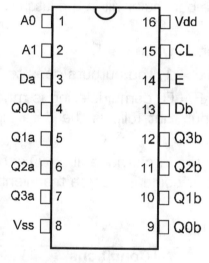

(Figure 2.94)

Pin Names:

Vdd – Positive Supply Voltage (3V to 15V)
Vss – Ground
CL – Clear Input
A0, A1 – Address Inputs
Q0 to Q3 – Outputs
D – Data Input
E – Enable

CMOS Sourcebook

Truth Table:

		Mode Selection		
E	CL	Addressed Latch	Unaddressed Latch	Mode
0	0	Follows Data	Holds Previous Data	Addressable Latch
0	1	Hold Previous Data	Holds Previous Data	Memory
1	0	Follows Data	Reset to "0"	Demultiplexer
1	1	Reset to "0"	Reset to "0"	Clear

Operation Mode:

- Data entered into a particular bit in the latch when addressed by the address inputs and E = 0.
- Data entry is inhibited when E = 0.
- When CL = 1 and E = 1, the outputs goes to "0".
- When CL = 1 and E = 0, demultiplexing is made: the bit addressed has an active output that follows the data input. The bits not addressed are held to "0".
- When operating in the latch mode, if E = 0 and CL = 0 are changing more than one bit, it could impose a transient wrong address.

Electrical Characteristics:

Characteristic	Conditions (Vdd)	Value	Units
Drain/Source Current	5V	0.88	mA
	10V	2.25	mA
	15V	8.8	mA
Propagation Delay Timer (Data to Output) (typ)	5V	200	ns
	10V	75	ns
	15V	50	ns
Quiescent Device Current (max)	5V	5	µA
	10V	10	µA
	15V	20	µA
Supply Voltage Range	3 to 15		V

Functional Diagrams and Information for Designers

Other Devices:
- The 4724 is an 8-bit addressable latch.

Applications:
- Multiline Decoders
- A/D Converters

Observations:
The outputs can drive 1 LS TTL input.

4724

8-Bit Addressable Latch

Description: This device is an 8-bit addressable latch with three address inputs, an active low-enable input, clear input, and eight outputs.

Functional Diagram or/and Package:

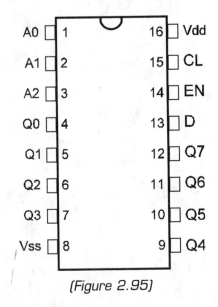

(Figure 2.95)

Pin Names:

Vdd – Positive Supply Voltage (3V to 15V)
Vss – Ground
CL – Clear
A0, A1, A2 – Address Inputs
Q0 to Q4 – Outputs
D – Data
EN – Enable

Functional Diagrams and Information for Designers

Truth Table:

EN	CL	Mode Selection		Mode
		Addressed Latch	Unaddressed Latch	
0	0	Follows Data	Holds Previous Data	Addressable Latch
0	1	Hold Previous Data	Holds Previous Data	Memory
1	0	Follows Data	Reset to "0"	Demultiplexer
1	1	Reset to "0"	Reset to "0"	Clear

Operation Mode:

- Data enter into a particular bit in the latch when addressed by the address inputs and EN = 0.
- Data entry is inhibited if EN = 1.
- When CL = 1 and EN = 1, the outputs goes to "0".
- When CL = 1 and EN = 0, demultiplexing occurs. The bit addressed has an active output that follows the data input while the other remains low.

Electrical Characteristics:

Characteristic	Conditions (Vdd)	Value	Units
Drain/Source Current	5V	0.88	mA
	10V	2.25	mA
	15V	8.8	mA
Propagation Delay Time (Data to Out) (Typ)	5V	200	ns
	10V	75	ns
	15V	50	ns
Quiescent Device Current (max)	5V	5	µA
	10V	10	µA
	15V	20	µA
Supply Voltage Range	3 to 15		V

Other Devices:
- The 4723 is a dual 4-bit addressable latch.

Applications:
- A/D Converters
- Multiline Decoders

Observations:
The outputs can drive one LS TTL input.

SECTION 3
BASIC BLOCKS USING CMOS ICs

Basic Blocks Using CMOS ICs

In the third part of the *CMOS Sourcebook* is found a large collection of basic circuits that can be used as part of more complex designs, or they can stand alone as complete devices. The circuits are not critical and parameter changes (external components and voltages) can alter the performance in a wide range of values. This means that it is very easy to make changes in the basic circuits to achieve the desired performance as an application may call for it. The basic blocks are divided in the next groups:

- Oscillators (Astables)
- Monostables
- Bistables and Counters
- Complete Applications

It's important to remember that the speed of the CMOS ICs depends on the sourced voltage, and in many applications where high-frequency signals are used, the designer must take this fact into account.

Oscillators

Due the tolerance of the components, changes in the suggested values often can be made to achieve the best performance of each circuit. Oscillators can be made using CMOS inverters combined with other functions. In the specific case of inverters, this function can be found in the 4049 or 4069 IC, but we also can build an inverter from other functions. OR, AND, NOR, and NAND gates can be used as inverters in the circuits suggested here.

Figure 3.1 shows how we can connect the remaining inputs of these functions to use them as inverters.

(Figure 3.1)

Although the basic inverters are represented in the applications, you are free to alter the configurations using an equivalent function adapted to the task.

Transistor Oscillator

This first basic circuit doesn't use CMOS functions, but instead it uses a bipolar transistor. The main feature for this circuit is its ability to drive a CMOS input. In the basic application shown in Figure 3.2, the circuit produces a 1 MHz signal, but the crystal can be changed in a wide band of values. However, remember that CMOS ICs have frequency limits based on the sourced voltage.

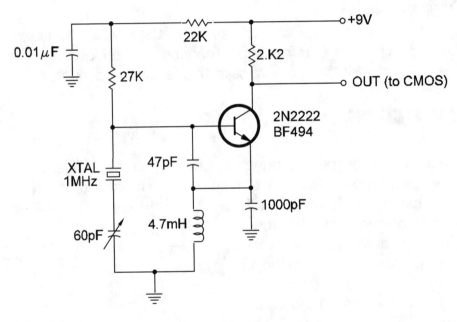

[Figure 3.2]

The capacitors must be ceramic types and the inductance may be altered if other frequencies are produced. The transistor is the BF494, but any general-purpose silicon transistor will function in this circuit, such as the 2N2222, BF495, etc.

Basic Blocks Using CMOS ICs

CMOS 555

The TLC555 is the CMOS equivalent of the well-known 555 bipolar transistor. It can be used in the production of signals in a range from a fraction of hertz to about 2 MHz. The frequency depends on Ra, Rb, and C, based on the formula in Figure 3.3, where the basic configuration is shown.

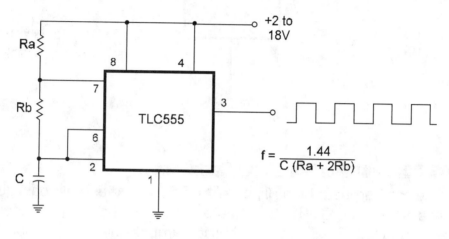

(Figure 3.3)

The common 555 (bipolar transistor) can also be used to drive CMOS circuits without any problem. The only difference is that the 555 drains more current than the CMOS equivalent. In this circuit, the lowest value for Ra and Rb is 1 k ohm and the lower value for C is 47 pF. The highest values are determined only by losses in the capacitor. The TLC555 will operate from power sources between 2 V and 18 V.

Gated CMOS 555

The CMOS TLC555 can be gated from external CMOS circuits using the configuration shown in Figure 3.4. The frequency of the signal when the control input goes to the high level is given by Ra, Rb and C, the same way as described in circuit 2. The oscillator is on when the control input is in the "1" logic level.

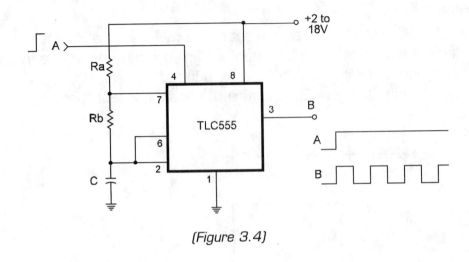

(Figure 3.4)

CMOS 555 with Variable Duty Cycle

The time the output is high (1) of a TLC555 in the astable configuration depends on Ra and C. The time the output remains low (0) depends on Ra, Rb, and C. It is easy to see that it is impossible to have duty cycles lower than 50 percent using this basic configuration. The circuit shown in Figure 3.5 is a variable-duty cycle oscillator that uses the TLC555 (and is also valid for the bipolar 555).

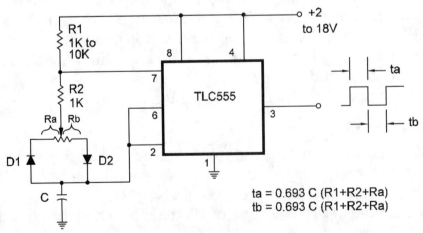

ta = 0.693 C (R1+R2+Ra)
tb = 0.693 C (R1+R2+Ra)

(Figure 3.5)

Basic Blocks Using CMOS ICs

The formulas to determine the frequency and duty cycle are given in the diagram. The diodes are general-purpose types, such as the 1N914 or 1N4148. The minimum values for the components are the same as in circuit 2.

Schmitt Trigger Oscillators

The advantages of using Schmitt Inverters (74106) or Schmitt NAND gates (4093) in oscillators lie in the need for a low number of external components and the snap action due to hysteresis that allows the production of high-quality square signals. The circuit shown in Figure 3.6 is the basic configuration using one of the four gates of a 4093 and one of the six inverters of a 40106. The frequency is given by the time constant RC according to the formula in the diagram.

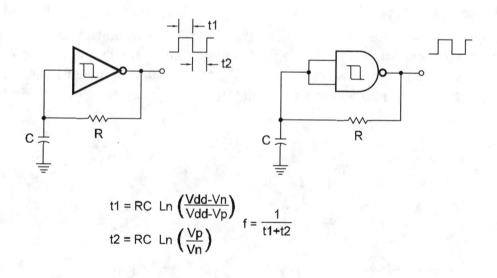

$$t1 = RC \ \text{Ln} \left(\frac{Vdd-Vn}{Vdd-Vp} \right)$$

$$t2 = RC \ \text{Ln} \left(\frac{Vp}{Vn} \right)$$

$$f = \frac{1}{t1+t2}$$

(Figure 3.6)

CMOS Sourcebook

In the formula:

- t is the total period in seconds (s)
- t1 is the period to the output in the high level in seconds (s)
- t2 is the period for the output in the low level in seconds (s)
- R is the resistance in ohms (O)
- C is the capacitance in farads (F)
- Vdd is the power-supply voltage in volts (V)
- Vp is the positive threshold voltage in volts (V)
- Vn is the negative threshold in volts (V)

The resistor can be replaced by a trimmer potentiometer to perform as a variable oscillator. A 100 k ohm trimmer potentiometer in series with a 10 k ohm resistor is suggested for a practical application in the audio range. The lowest value recommended for C is 47 pF and for R is 1 k ohm. The highest value of the components depends on the losses of C.

Photo Oscillator

Replacing the resistor for an LDR (light dependent resistor or CDs cell) makes the oscillator independent of the incident. The frequency depends on C. The circuit shown in Figure 3.7 uses a NAND gate of a 4093 or any Schmitt Inverter, such as the 40106.

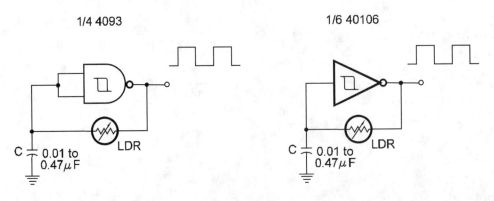

(Figure 3.7)

Basic Blocks Using CMOS ICs

Common LDRs usually have a resistance of some tenths or hundreds ohm in natural light and high as some megohm in complete darkness. In an oscillator such as the one suggested in Figure 3.7, the frequency range is as wide as 1000:1. Operation in the audio range will be achieved with capacitors in a range between 0.01 uF and 0.47 uF.

LC Oscillator Using the 4093

A high-frequency square wave oscillator using an LC-tuned circuit is shown in Figure 3.8. The upper frequency limit for this configuration is about 6 MHz when powered from 10 V supplies. The lowest value recommended for C is 47 pF. Due to the high-frequency operation, the output isn't a perfect square signal.

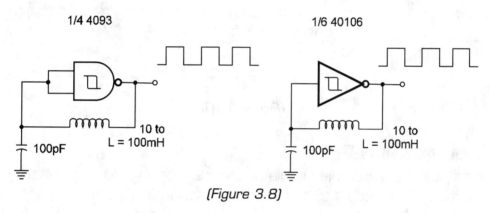

(Figure 3.8)

LC Oscillator Using the 4069

Another oscillator that uses an LC network is shown in Figure 3.9. The upper limit depends on the power-supply voltage, which is about 4 MHz for 10 V. Notice that the 4069 is formed by simple inverters.

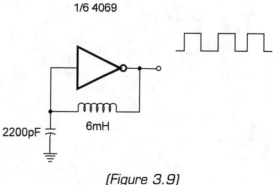

(Figure 3.9)

Gated Oscillator

The circuit shown in Figure 3.10 is gated by an external signal. A high logic level applied to the input starts the oscillation. The frequency depends on the RC network as described in circuit 7. Waveforms for this circuit are also shown in the figure. The maximum frequency is about 4 MHz and depends on the power-supply voltage.

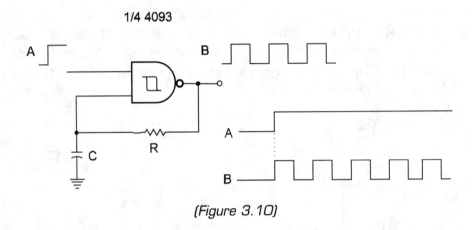

(Figure 3.10)

Programmed Duty Cycle

The duty cycle of a Schmitt Oscillator can be programmed by R1 and R2 in this circuit. The frequency of the circuit shown in Figure 3.11 is determined by the two resistors and C. D1 and D2 are general-purpose silicon diodes, such as the 1N914 or 1N4148. The time the output is "1" depends on R1 and C. The time the output is "0" depends on R2 and C. The circuit can be based on the Schmitt NAND gates of a 4093 or inverters of a 40106. The formulas for the periods given with circuit 5 can be used to calculate T1 using R1 and T2 using R2.

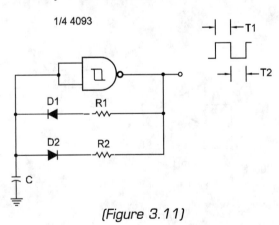

(Figure 3.11)

Basic Blocks Using CMOS ICs

Crystal NAND Oscillator

The circuit shown in Figure 3.12 can produce frequencies from 100 kHz to 2 MHz. The inverter can also be replaced by equivalents like those shown in the beginning of this section. This circuit is suitable for applications, such as clocks and precision signal generators.

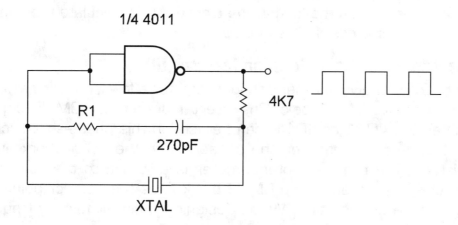

(Figure 3.12)

Sine Wave Oscillator Using an Inverter (I)

The circuit shown by Figure 3.13 can be used to produce low-frequency sine wave signals from CMOS logic. The frequency depends on the Twin-T network and is calculated by the formula given with the diagram. The ratio of values between the capacitors and resistors

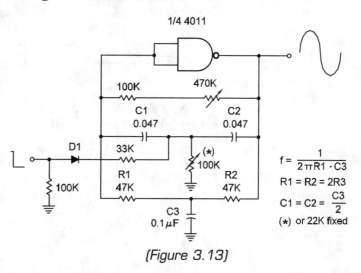

(Figure 3.13)

$$f = \frac{1}{2\pi R1 \cdot C3}$$

R1 = R2 = 2R3

$$C1 = C2 = \frac{C3}{2}$$

(∗) or 22K fixed

CMOS Sourcebook

may be observed. The circuit is controlled by an external logic signal applied to the input. The input resistor (27 k ohm) can be changed to modify the damp effect of the control signal. Damped signals can also be adjusted replacing the 100 k ohm resistor between the capacitor in the twin-T network by a trimmer potentiometer.

When designing a circuit to any other frequency, keep the resistors with the values as recommended in the diagram and calculate the capacitor according to the desired frequency.

Sine Wave Oscillator Using an Inverter (II)

Figure 3.14 shows another configuration for a low-frequency sine wave oscillator using CMOS logic. One inverter of common CMOS functions, such as the 4001 or 4011, can be used in this phase-shift oscillator. The frequency depends on the capacitors in the RC network and the resistors. The trimmer potentiometer is adjusted to the point where the voltage supplied to the CMOS block finds the correct point in the linear mode of operation. When designing a circuit for any frequency,

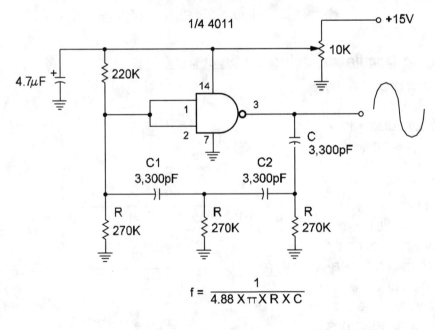

$$f = \frac{1}{4.88 \times \pi \times R \times C}$$

(Figure 3.14)

keep the resistors with the values as recommended in this block and calculate the capacitors. This circuit is suitable for audio signals in the band between a low hertz and 10 kHz. When calculating the components for a new frequency, keep the resistors with the values as suggested and calculate C.

Two-Inverter Oscillator (I)

Figure 3.15 shows the basic configuration of a ring oscillator using CMOS logic. The frequency depends on R1 and C1. R1 must be in the range between 10 k ohm and 22 M ohm. The lowest value recommended for C1 is 47 pF. The highest frequency produced by this circuit depends on the power-supply voltage, but it is generally about 6 megahertz. Any inverter, such as the ones suggested earlier in this section, will operate in this circuit. The figure also shows the waveforms in the main points of the circuit. When producing low-frequency signals, Schmitt Trigger devices are recommended.

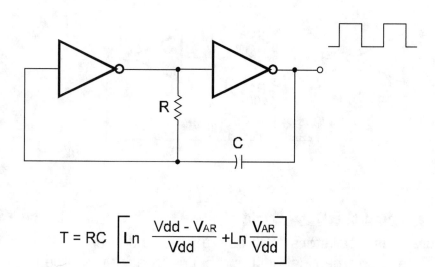

$$T = RC \left[Ln \frac{Vdd - V_{AR}}{Vdd} + Ln \frac{V_{AR}}{Vdd} \right]$$

(Figure 3.15)

In the formula:

- T is the period in seconds (s)
- R is the resistance in ohms (Ω)
- C is the capacitance in Farads
- Vdd is the power-supply voltage (V)
- Vtr is the transfer voltage (V) – see Section 2

Improved Two-Inverter Oscillator

The simplest version of a ring oscillator using two inverters like the one shown in Figure 3.15 has some problems. It is dependent on the voltage source and some deformations can be observed in the output signal — it is not perfectly square. The best performance of a ring oscillator using two inverters can be achieved with the configuration shown in Figure 3.16. The added resistor avoids the RC network to be loaded by the clamping diodes inside the CMOS IC. The resistors and C1 determine the frequency.

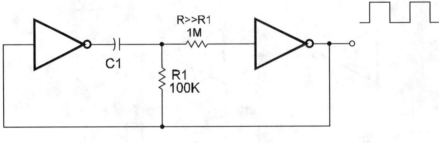

(Figure 3.16)

Programming the Duty Cycle

The previous oscillators that use gates, and the 4093 in the basic version, are symmetrical because 50 percent of the time the output is at the "1" logic level, and the other 50 percent at the "0" logic level. We say that they are square wave generators or the duty cycle is 50 percent. If different duty cycles are needed, you can use other configurations. The duty cycle for this oscillator depends on the resistors R1 and

Basic Blocks Using CMOS ICs

R2. The mark depends on R1. The space depends on R2. The diode is any general-purpose unit, such as the 1N914 or 1N4148. With the values as shown in Figure 3.17, operation frequency will be in the audio range.

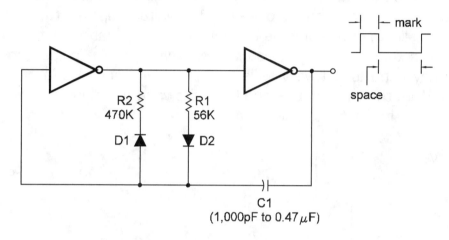

(Figure 3.17)

Variable Duty Cycle

Figure 3.18 shows another circuit that allows adjustments in the duty cycle of the output signal. The position of the cursor of the potentiometer acts on the duty cycle with few changes in the frequency. The diodes are general-purpose types. With the values as shown in the figure, the oscillator will operate in the audio range. The logic functions are inverters or other gates that can be used as inverters. Using the formula in Figure 3.16, the complete period can be divided in two parts (T1 and T2). Apply the formula for T1 using R1, and for T2 using R2 to calculate the duty cycle and/or the frequency.

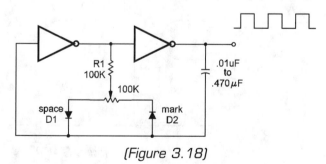

(Figure 3.18)

Voltage Controlled Oscillator (VCO)

A simple VCO using two inverters in a ring configuration is shown in Figure 3.19. A VCO is a circuit in which the frequency depends on the voltage applied to the input. It can be used in applications, such as musical instruments and sirens. The central frequency depends on R1 and C1. This frequency changes based on the voltage applied to the control input. The limits of this voltage depend on the application, but the circuit can be used as an FM generator in applications where the generated frequency band is not very wide. For a wider band, VCOs prefer the 4046. With the values for R and C as shown in the diagram, the circuit will oscillate in the audio band.

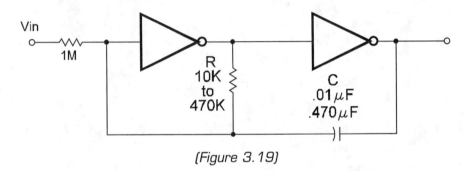

(Figure 3.19)

Gated Oscillator (I)

The gated oscillator shown in Figure 3.20 turns on when a "1" logic level is applied to the input. The frequency depends on C1 and R1. With the values shown in the figure, the circuit will operate in the audio

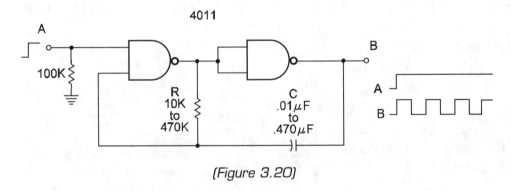

(Figure 3.20)

range. Any AND function can be used in this circuit. The 100 k ohm resistor in the input is necessary if, when the circuit is not gated, the control input is open.

Gated Oscillator (II)

The gated oscillator in Figure 3.21 turns on when a "0" logic level is applied to its input. The frequency, as in the previous device, depends on R1 and C1. Using the values in the figure, the circuit will operate in the audio range. Any NOR gate can be used in the configuration.

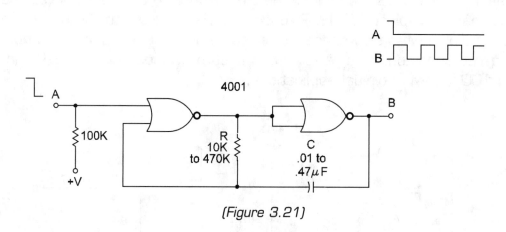

(Figure 3.21)

Three-Gate Ring Oscillator

An oscillator that uses two gates is not suitable for critical applications, as it tends to pick up noise from the power-supply line. This noise can affect the output, producing deformed signals. A better type for these applications is the three-gate ring oscillator. It uses three inverters instead of only two. Figure 3.22 shows the

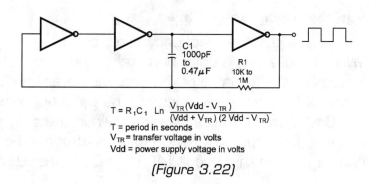

$$T = R_1 C_1 \ \text{Ln} \ \frac{V_{TR}(V_{dd} - V_{TR})}{(V_{dd} + V_{TR})(2 V_{dd} - V_{TR})}$$

T = period in seconds
V_{TR} = transfer voltage in volts
V_{dd} = power supply voltage in volts

(Figure 3.22)

CMOS Sourcebook

basic configuration for a three-gate ring oscillator using inverters (or any other function wired as inverter). The frequency depends on R1 and C1. Using the values shown in the figure, the circuit will operate in the audio range. The lowest value recommended for C1 is 47 pF and is 10 k ohm for R1. The frequency for this circuit can be calculated by using the formula in the diagram.

Gated Oscillator – Three Gates

To turn on the previous configuration using a "1" logic level, the circuit is given in Figure 3.23. To trigger the circuit on with a "0" logic level, the gates can be replaced by NOR functions. The resistor to the ground in the control input is only necessary if the input remains open in the non-triggered condition. This resistor can assume values between 1 k ohm and 100 k ohm in typical applications.

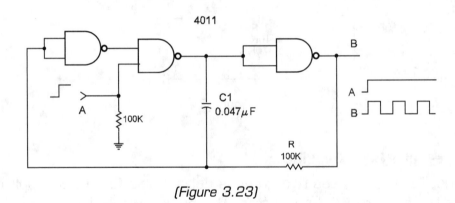

(Figure 3.23)

Variable Oscillator – Three Gates

You can adjust the frequency produced by a ring oscillator with three inverters inserting a potentiometer in the feedback network, as shown in Figure 3.24. The value of the potentiometer can be increased if a wider range of frequencies is needed. The potentiometer can adopt values up to 10 M ohm. With the values shown in the diagram, the signals will be in the audio range. Any CMOS function that can be connected as inverter can be used in this circuit (4011, 4001, 4069, etc.).

Basic Blocks Using CMOS ICs

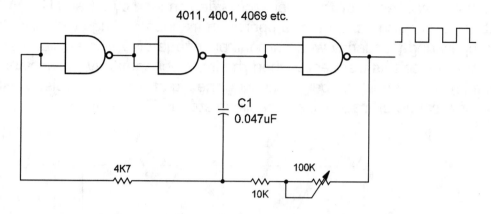

(Figure 3.24)

Voltage-Controlled Oscillator (VCO) Using the 4046

A VCO or VCF (Voltage to Frequency Converter) can be built using part of a 4046 (PLL) circuit. The basic circuit is shown in Figure 3.25. This circuit can be used to produce signals in a frequency range of 100:1 to 1000:1 cf based on the components used as reference. The configuration uses the VCO existing inside the 4046 that can operate in frequencies up to 1.6 MHz when powered from 15 V supplies (see data in Section 2 for more information).

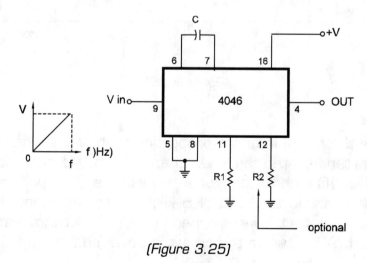

(Figure 3.25)

CMOS Sourcebook

The basic frequency for this oscillator is determined by C1 and R1. Without any other components, the operation frequency range will vary from the operating frequency when a control voltage of 0 V is applied to the input, to twice this frequency when the input voltage is +V (power-supply voltage). This range can be shifted with the aid of a second resistor (R2). The new range of frequencies are illustrated in Figure 3.26.

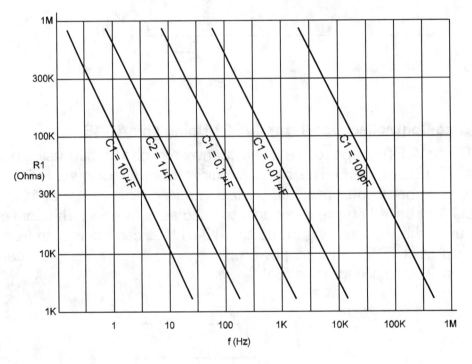

(Figure 3.26)

R2 will raise the range in direct proportion to its value when compared to R1. For instance, if the range will change from 0 kHz to 1 kHz without R2, and R1 is 10 k ohm, then when inserting a 20 k ohm for R2, the range will be changed to 0.5 to 1.5 kHz. Other important characteristics for this circuit are the input impedance of 10^6 M ohm and operating current of only 90 uA when the circuit is powered from a 10 V source.

Basic Blocks Using CMOS ICs

Simple Voltage-Controlled Oscillator (VCO) Using the 4046

An application for the previous circuit is shown in Figure 3.27. The central frequency is determined by C and R1. The potentiometer is used to control the range. The minimum value recommended for C1 is 47 pF. Variations in this circuit include the use of a resistive network with a sensor (an LDR for instance). In this case, the frequency will be dependent on the amount of light in the sensor. Another type of sensor that can be used is an NTC. The frequency in this case will depend on the temperature. The linear characteristic of the VCO makes it suitable for applications in instrumentation. Digital thermometers or photometers can be designed from this circuit.

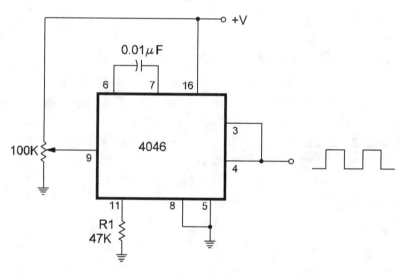

(Figure 3.27)

Two Frequencies Voltage-Controlled Oscillator (VCO)

The circuit shown in Figure 3.28 uses a flip-flop found inside the 4046 to divide the original frequency produced by the VCO by two. This allows the designer to get two signals with different frequencies: the original and the frequency divided by two. The operation principle is the same as the circuit in Figure 3.24 and the frequency can be calculated as described there.

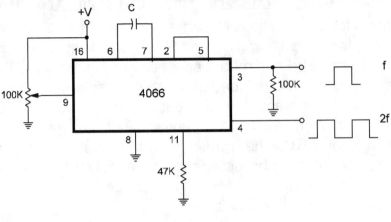

(Figure 3.28)

One-Component Oscillator

Figure 3.29 shows a ring oscillator using five inverters of a 4069. (The circuit operates only with an odd number of gates.) The frequency depends on the time delay of the signal when passing through each gate. Because this time depends on the power-supply voltage, a wide range of frequencies can be adjusted by the voltage divider in the power-supply input. With a power supply of 6 V, the circuit will generate signals typically between 1 MHz and 4 MHz.

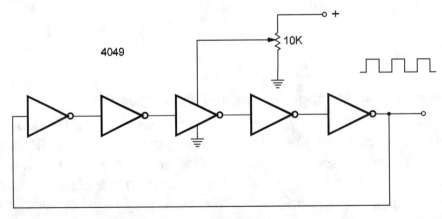

(Figure 3.29)

Basic Blocks Using CMOS ICs

Variable Frequency Ring Oscillator

Another way to change the frequency of the previous oscillator is to add a variable capacitor in the feedback loop as shown in Figure 3.30. This capacitor can typically be in a range from 2-20 to 5-100 pF. The upper frequency limit for this oscillator is about 4 MHz with a 12 V power supply. The upper limit of frequency depends on the times the signal passes across each gate.

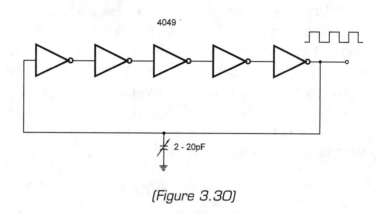

(Figure 3.30)

Light-Controlled Oscillator (I)

This circuit uses a phototransistor as sensor. The changes in the resistance of the transistor, based on the light falling onto it, alter the mark of the signal and also the final frequency. Any phototransistor can be used in the circuit shown in Figure 3.31. It is also possible to use an LDR. The capacitor and R1 determine the central frequency of the circuit. With the values given in the diagram, the signal will be in the audio band (1 kHz to 10 kHz).

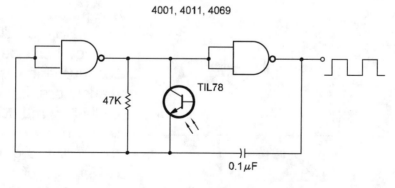

(Figure 3.31)

Light-Controlled Oscillator (II)

Light falling on the transistor changes both the mark and the space of the generated signal, thanks to the diode bridge. The diodes are general-purpose types, such as the 1N4148 and 1N914. Any phototransistor can be used in this circuit. The central frequency depends on C and the resistance presented by the transistor in a reference light level. The circuit shown in Figure 3.32 can use any inverter. For signals in the audio band, C can assume values between 4,700 pF and 0.47 uF. The lowest value recommended for C is 47 pF.

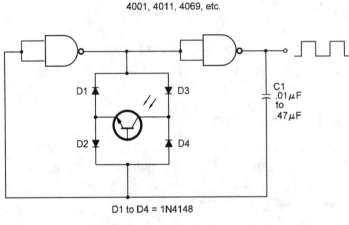

(Figure 3.32)

Current-Controlled Oscillator (CCO)

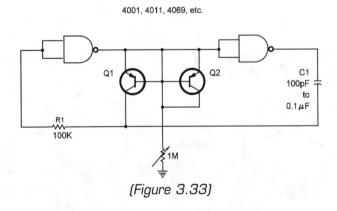

(Figure 3.33)

The circuit shown in Figure 3.33 is a CCO. The frequency here depends on the current through the transistors. It can be controlled by a 1 M ohm potentiometer. The central frequency is given by the resistance of the transistor in a reference level and

C1. The range of values for R1 is between 1 k ohm and 100 k ohm. For C1 the recommended band is between 100 pF and 100 nF. With the values shown in the diagram, the circuit will operate in the audio band.

Two-Phase Oscillator

The circuit shown in Figure 3.34 uses a 4047 to produce three signals whose frequencies are determined by C and R, based on the formula in the schematic diagram. The 4047 is a Monostable/Astable Multivibrator with a frequency divider and symmetrical outputs.

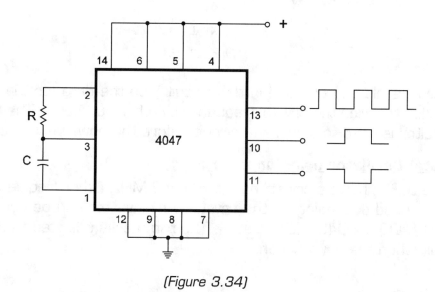

(Figure 3.34)

One of the signals has a frequency determined by R and C, and the two others have opposite phases which are half of the central frequency. The minimum value recommended for R is 1 k ohm and for C is 10 pF. Using a 22 k ohm resistor and a 10-pF capacitor, the circuit will run at 1 MHz. For 220 k ohm and 1000 pF, the circuit will run at 1 kHz.

200 kHz Oscillator

A precision 200 kHz oscillator using a 3.2768 kHz crystal can be built with a 4060 IC as shown in Figure 3.35. The frequency is divided by the sequence of flip-flops in the 4060, and a 200 kHz signal is the output.

CMOS Sourcebook

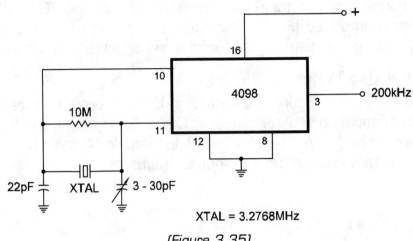

(Figure 3.35)

Other crystals can be used in this circuit (see the data for the 4060 if you want to divide the crystal frequency by other numbers). The trimmer capacitor is used to adjust the circuit before the power is turned on.

Crystal Oscillator using an Inverter

Figure 3.36 shows a crystal oscillator for 2 MHz. Other frequencies can be produced depending on the crystal. Any inverter can be used in this circuit (4001, 4049, 4011, etc.). The trimmer is adjusted to start the circuit when the power is on.

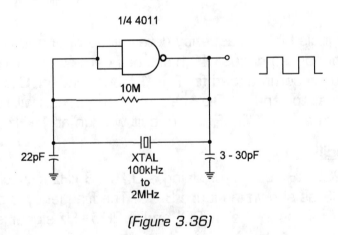

(Figure 3.36)

Monostables (One Shot)

Monostables are circuits that can be turned by an external pulse, remaining on during a time interval determined by an external network. Generally RC networks are used to determine the time the output is on or the time delay of a monostable. Monostable circuits, or "one shot" as they are also called, can be used as timers, contact debouncers, pulse conformation, time delay circuits, missing pulse detectors, and for many other applications.

The following are basic monostable circuits suitable for use with CMOS circuits and CMOS monostables. The external components can be changed in a large range of values depending on the application and compensation of the device tolerances.

Basic Monostable Using the 7555 (CMOS)

The CMOS TLC555 (equivalent to the bipolar 555) can be used as a monostable in the configuration shown in Figure 3.37. The time the output remains high depends on R1 and C1, based on the formula in the diagram. The minimum value recommended for R1 is 1 k ohm and for C1 is 100 pF. The highest value for these components depends on the

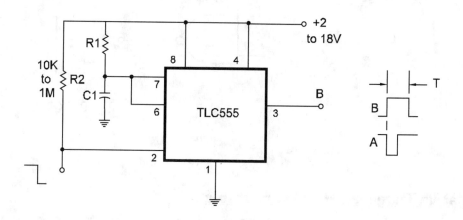

(Figure 3.37)

CMOS Sourcebook

losses in the capacitor and is typically 2.2 M ohm for R1 and 1500 uF for C1. The TLC555 can drive or source 200 mA in its output — directly driving relays and lamps.

This circuit is triggered when the input (pin 2) is grounded. When waiting, pin 2 may be held at the high logic level. This can be done with a resistor between 10 k ohm and 1 M ohm.

In the formula:

- T is the period in seconds (s)
- R is the resistance in ohms (O)
- C is the capacitance in farads (F)

Positive-Triggered Half Monostable

One 40106 Schmitt inverter or one 4093 gate can be used as a half monostable to be triggered by the positive transition of the input. The amount of time the circuit is on is approximately 0.7 x R x C (R in ohm and C in farad). It is the circuit shown in Figure 3.38. The minimum value recommended for C is about 100 pF. For long monostable action, the TLC555 or the bipolar 555 is recommended.

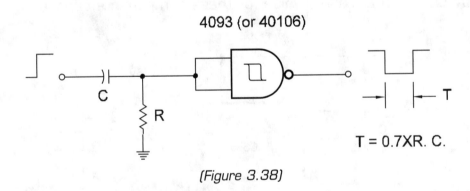

(Figure 3.38)

Negative-Triggered Half Monostable

This circuit is the equivalent of the previous one, but it is triggered with the negative transition of the input pulse. The amount of time the output

Basic Blocks Using CMOS ICs

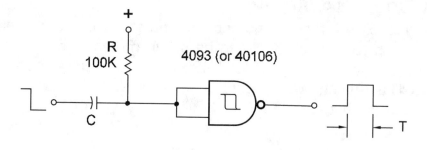

$$T = 0.7 \times R \cdot C$$

(Figure 3.39)

is activated is 0.7 x R x C (R in ohm and C in farad). The circuit is shown in Figure 3.39. The minimum recommended value for C is about 100 pF.

Delayed Turn-On Gate

This circuit can be used to add a delay to a logic signal, as shown by the input and output waveforms in Figure 3.40. The delay is given by R x C and is the same as the rise time and fall time of the output when compared to the input. The circuit will also function with common NAND gates.

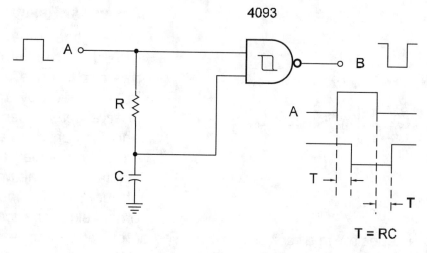

(Figure 3.40)

Delayed Turn-On and Off Gate

Different delays for the rise time and fall time of the output signal when compared with the input signal can be programmed using this circuit. The delays are given by R1 x C1 and R2 x C2. The circuit uses a gate of the 4093 IC and is shown in Figure 3.41 along with the corresponding waveforms in the output and input.

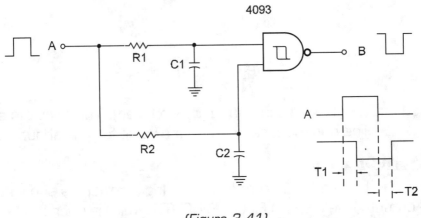

(Figure 3.41)

Power-On Reset

When the power is turned on, the output of this circuit goes to the "1" logic level by a time interval determined by R and C. For the values shown in the diagram of Figure 3.42, the time is about 1.4 seconds. If other values are needed, change the capacitor in direct proportion. The output pulse can be used to reset a function in a circuit or to drive a relay.

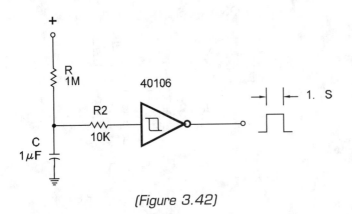

(Figure 3.42)

Two NAND Gates Monostable

Two NAND gates are used in the configuration shown in Figure 3.43 to perform as a monostable triggered by the positive edge of the input signal. The time the output of the circuit remains "1" is about 0.7 x R x C (R in ohm and C in farad). Other NAND gates can be used in the same configuration. If the input remains unconnected to any circuit when triggered off, the addition of a ground resistor is recommended to prevent unstable operation. The value may be in the range between 10 k ohm and 10 M ohm. A suitable value if the circuit is triggered from CMOS logic is 100 k ohm.

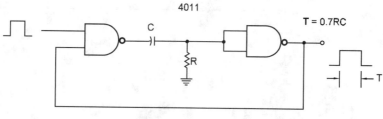

(Figure 3.43)

Two NOR Gates Monostable

To trigger the monostable shown in Figure 3.44 using the negative edge of a pulse, NOR gates are used. The delay is 0.7 x R x C (R in ohm and C in farad). If the input remains unconnected to any circuit, add a pull-up resistor between 10 k ohm and 10 M ohm, depending on the application, to prevent erratic behavior. A recommended value for general applications is 100 k ohm.

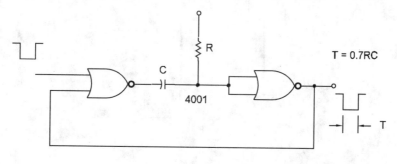

(Figure 3.44)

Adding Reset

Figure 3.45 shows how to add a reset key to the monostable using two NAND gates. When the key is pressed, timing is interrupted. Only when a new pulse is applied to the input is the monostable triggered again. The same procedure is valid for the NOR monostable connecting the key to ground.

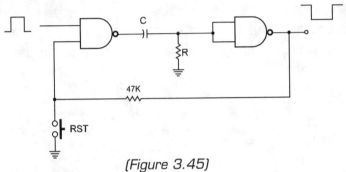

(Figure 3.45)

Monostable Using the 4013 Flip-Flop

One of the two D-type Flip-Flops found in a 4013 can be used as a non-retriggerable monostable, as shown in Figure 3.46. This circuit is triggered by a positive pulse applied to the input. When this pulse is applied the output, Q goes to the high logic level and Q/ goes to the low logic level as determined by a time delay fixed by R and C. The time constant for this circuit is about $0.7 \times R \times C$ (R in ohm and C in farad). The minimum recommended value for C is about 47 pF.

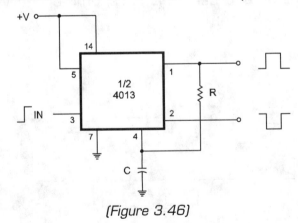

(Figure 3.46)

Basic Blocks Using CMOS ICs

Improved Monostable Using the 4013 Flip-Flop

An improved version of the previous circuit is shown in Figure 3.47. The resistor R1 must be at least five times larger than R. The diode is any general-purpose silicon type, such as the 1N4148 or 1N914. This circuit is also triggered by the positive transition of the input signal. The minimum value recommended for C is 47 pF. The outputs are complementary to each other.

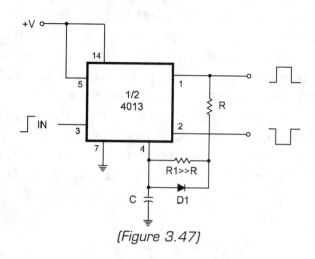

(Figure 3.47)

Monostable Using the 4027

A configuration like the one used with the 4013 is also suitable for the 4027. The amount of time the output is "1" after the circuit is triggered is about $0.7 \times R \times C$. R1 must be at least five times larger than R in this circuit. The diode is any general-purpose type, such as the 1N4148 or 1N914. The circuit is shown in Figure 3.48. It is triggered by the positive transition of the input signal. The 4027 is a Dual J-K Master/Slave Flip-Flop with Set and Reset. For more information about this device, see Section 2.

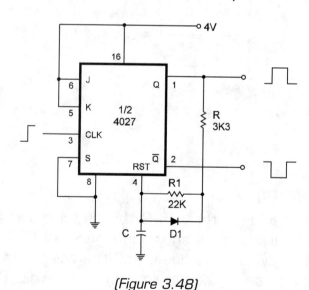

(Figure 3.48)

Monostable Using the 4528 (I)

The 4528 is a dual-retriggerable monostable that needs only the timing network formed by R and C to perform this function, as shown in Figure 3.49. This circuit is triggered on with the positive edge of the input signal. The minimum value recommended for C is 100 pF and R can assume values between 10 k ohm and 10 M ohm. Time constant is about R x C. This circuit is not recommended for large time delays; use the 555 CMOS or another. The numbers of pins used in this circuit correspond to one of the two monostable multivibrators of the 4028 package. To use the other monstable, see Section 2 for more information.

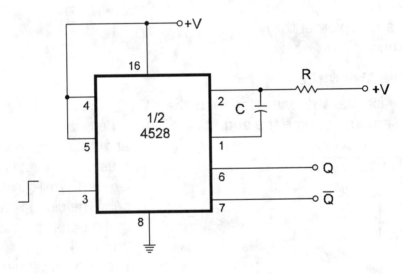

[Figure 3.49]

Monostable Using the 4528 (II)

A version of a monostable that uses the 4528, but is triggered by negative transitions of the input signal, is shown in Figure 3.50. This circuit operates just as the previous one does, and the resistor and capacitor have the same limits as in the previous case as well. The time constant is also given by these two components and is approxi-

Basic Blocks Using CMOS ICs

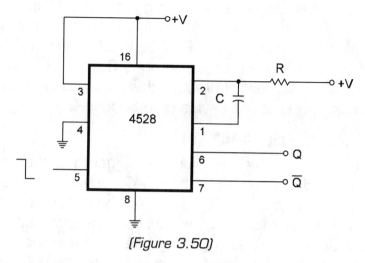

(Figure 3.50)

mately R x C. The numbers of pins correspond to one of the two monstables of the 4028 package. If you intend to use the other one, see Section 2 for the pin numbers.

Monostable Using the 4047 (I)

The 4047 is a monostable/astable CMOS multivibrator. The basic configuration is triggered by the positive transition of the input signal, and can be seen in Figure 3.51. The time constant for this circuit is 2.4 x R x C. The circuit has complementary outputs. When triggered, Q goes to

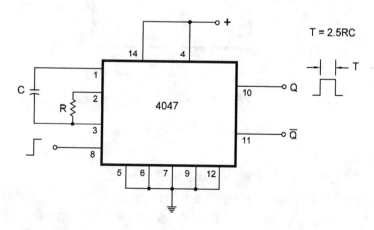

(Figure 3.51)

the "1" logic level at the same time Q/ goes to the "0" logic level. R must be in the range of 10 k ohm and 10 M ohm, and the maximum value recommended for C is 100 pF. If R is 22 k ohm and C is 10 pF, then the amount of time for this circuit to be on is 2 us. If R is 220 k and C is 1000 pF, then the amount of time for this circuit to be on is 550 us. For more information about the device, see Section 2.

Monostable Using the 4047 (II)

A version of a monostable multivibrator triggered by the negative transition of the input signal and using the 4047 is shown in Figure 3.52. The time constant is around 2.4 X R x C, and these components may be in the same range as recommended in the previous configuration. The amount of time on is the same if the values given in the previous circuit are used. Notice that the circuit has complementary outputs due the flip-flop found inside the package.

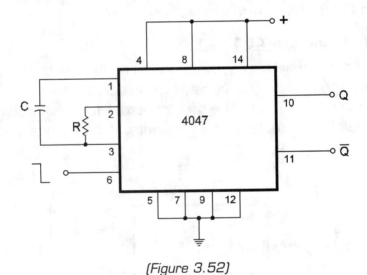

(Figure 3.52)

Retriggerable Monostable Using the 4047

The previously mentioned versions of monostable multivibrators are not retriggerable. In order to add this characteristic, some changes in the configuration must be made, as shown in Figure 3.53. The other char-

Basic Blocks Using CMOS ICs

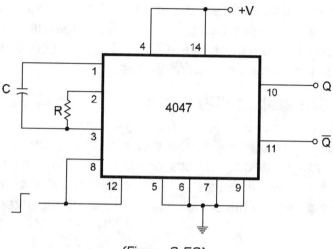

(Figure 3.53)

acteristics of the circuit remain the same, including being triggered by positive transitions of the input signal and the outputs having complementary signals.

Monostable Using the 4098 (I)

The 4098 IC contains two monostable multivibrators in the same package. The basic configuration for this IC is shown in Figure 3.54. The numbers in "brackets" represent the pins of the second multivibrator.

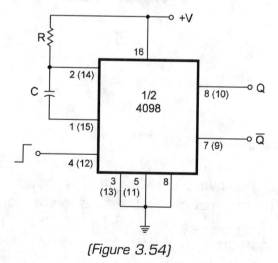

(Figure 3.54)

CMOS Sourcebook

This is a retriggerable and resettable multivibrator triggered by the positive transition of the input signal. The minimum value for R is 5 k ohm and the maximum capacitance is 100 uF. The time constant is 0.5 x R x C. The circuit has complementary outputs.

Monostable Using the 4098 (II)

A new version of the previous circuit triggered by the negative transition of the input signal is shown in Figure 3.55. The general characteristics described in the previous projects are the same for this configuration. The numbers between "brackets" correspond to the second multivibrator of the same package. This circuit also has complementary outputs.

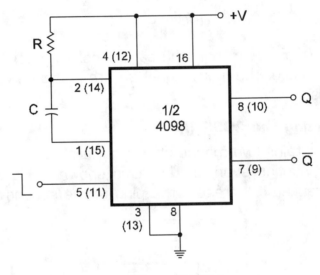

[Figure 3.55]

Monostable Using the 4098 (III)

The third version for the 4098 is shown in Figure 3.56. It is a monostable triggered by the positive transition of the input signal, but it is not retriggerable. The general characteristics for this configuration are the same as those of the circuit shown in Figure 3.52. This circuit also has complementary outputs.

Basic Blocks Using CMOS ICs

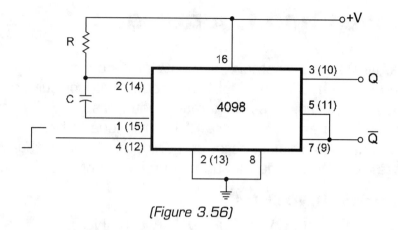

(Figure 3.56)

Monostable Using the 4098 (IV)

A fourth version of a non-retriggerable monostable is shown in Figure 3.57. It is triggered by the negative transition of the input signal. The time constant and the limit values for R and C are the same as described in circuit 52 (Figure 3.52). This circuit also has complementary outputs. The numbers in the diagram are for one of the two monostables found in this package. For the other monostable, see Section 2 for more information.

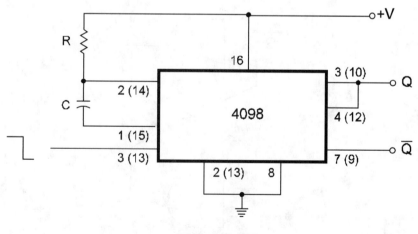

(Figure 3.57)

Bistables and Counters

Bistables, flip-flops, or memory elements are a category of devices and circuits that are used in many logic projects. These circuits can be used as frequency dividers, one-bit memory, counters, and for many other applications. The next circuits are CMOS flip-flops in basic applications, such as counters, shift registers, and dividers. Some of them use common gates and others, the functions found in complete devices.

Flip-Flop Using NAND Gates

The simplest flip-flop can be built with two NAND gates, as shown in Figure 3.58. This is a Reset-Set (S-R) type or R-S flip-flop (also called Set-Reset (S-R) in some publications). If the inputs are kept unconnected when the circuit is not triggered, the inputs must be connected to +V using pull-up resistors. Typical values for these resistors are 100 k ohm. This circuit is set and reset with negative pulses applied to the input. The same configuration can be built using two Schmitt NAND gates, such as the ones found in a 4093. Notice that set and reset are done with negative pulses applied to the inputs.

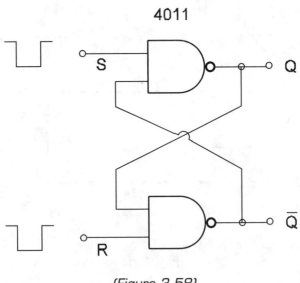

(Figure 3.58)

Basic Blocks Using CMOS ICs

Flip-Flop Using NOR Gates

The R-S flip-flop shown in Figure 3.59 uses two NOR gates and is set and reset by positive pulses applied to these inputs. The inputs must be kept in the low logic level when not triggered. This can be done with 100 k ohm connected to ground.

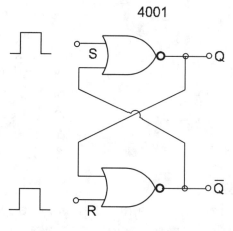

(Figure 3.59)

Touch-Triggered Flip-Flop

The 4093 or any other NAND gate can be used for the very sensitive configuration of a R-S flip-flop triggered by the touch of the finger in two sensors. The circuit is shown in Figure 3.60. The sensitivity depends on R1 and R2. The larger these resistors are, the higher the sensitivity. The sensors are two small metal plates that can be touched at the same time with the fingers.

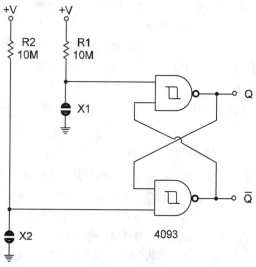

(Figure 3.60)

Clocked R-S Flip-Flop

The four NAND gates of a 4011 are used in the clocked R-S flip-flop shown in Figure 3.61. The circuit is set and reset by positive pulses in the inputs, and these pulses pass when the clock input is high. If the inputs remain open at moments during the operation, a 100 k ohm resistor to ground is needed. Other NAND gates can be used in the same configuration as the ones found in a 4093.

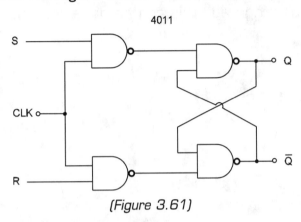

(Figure 3.61)

Alternate Action Switch

This circuit turns on and off when a short pulse is applied by SW to the input of the first inverter. The capacitor can assume values between 0.22 uF and 2.2 uF. Other gates that function as inverters can be used in this circuit, including the Schmitt types, such as the 4093. The pulses can be produced by sensors, such as reed switches or relays, in robotic and mechanical applications.

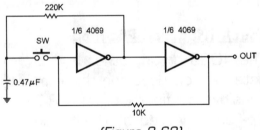

(Figure 3.62)

Touch Switch

The very high-input impedance of the CMOS gate allows this kind of device to be used as a sensitive touch switch, as shown in Figure 3.63. This circuit turns on and off with a single touch to the sensor. The 10 M ohm resistor determines the sensitivity and can be altered to function in the range between 1 M ohm and 22 M ohm. The higher the resistor, the better the sensitivity. The capacitor also can be changed as long as

Basic Blocks Using CMOS ICs

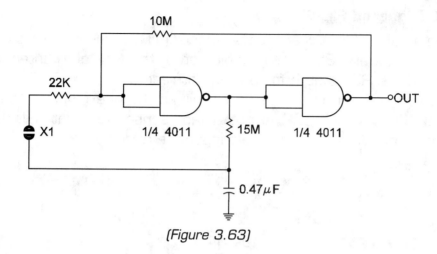

(Figure 3.63)

it stays in the range between 0.1 uF and 1 uF depending on the application. Any NAND gate can be used including the Schmitt types, such as the 4093.

Alternate Action Switch (II)

Another alternate action switch that uses a NAND Schmitt Gate (4093 IC) is shown in Figure 3.64. The triggering point for the best performance is adjusted in the trimmer potentiometer. The capacitor can be in the range between 0.47 uF and 2.2 uF.

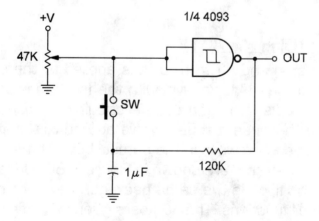

(Figure 3.64)

CMOS Sourcebook

Light-Triggered Flip-Flop

When a pulse of light reaches the sensor (a CDs cell or LDR), the flip-flop changes its state. Sensitivity is adjusted by the 1 M ohm potentiometer. Pressing the RST switch for a moment resets the circuit. Other NAND gates can be used in the same circuit as the one found in the 4093. Remember that the LDR is not a fast sensor for light pulses and a minimum duration is needed to trigger the circuit.

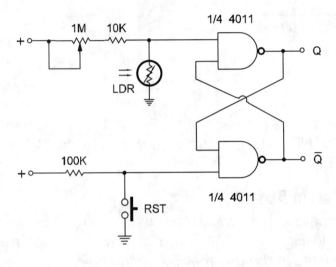

(Figure 3.65)

Basic Flip-Flop Using the 4013

This circuit turns on and off from pulses applied to the CLK input. The output Q goes to the "1" logic level with the first transition of the input impulse from positive to negative. The circuit can also be used as a frequency divider because a square wave applied to the input will correspond to another square wave in the output, but with half the frequency. The circuit and waveforms are shown in Figure 3.66. Because the 4013 has two D-type flip-flops in the same package, the other can be used in the same or other functions. The numbers given in the figure correspond to one of the flip-flops. For information on the other, see Section 2.

Basic Blocks Using CMOS ICs

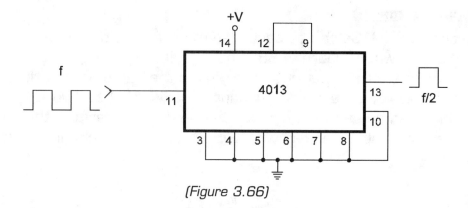

(Figure 3.66)

Basic Flip-Flop Using the 4027

The basic circuit of a flip-flop for use as a frequency divider or one-bit memory is shown in Figure 3.67. This circuit is based in the 4027, a Dual J-K Master/Slave Flip-Flop with Set and Reset. It can be used as a cell of a binary counter or divide-by-2 counter. The numbers in the diagram correspond to half of the package or one flip-flop. For the other half, see Part 2.

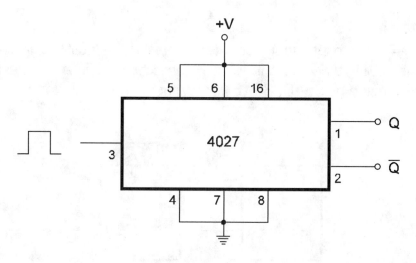

(Figure 3.67)

CMOS Sourcebook

Cascading the 4013

Two or more 4013s can be cascaded to perform a divide-by-4 (-8, -16, etc.), as shown in Figure 3.68. The circuit can be used as a two-bit binary counter. The input characteristic is the same as that described for circuit 64 (Figure 3.64). Because the 4013 has two D-type flip-flops in the same package, one device is enough to perform this circuit. The maximum input frequency is 12.5 MHz when the circuit is powered with 10 V.

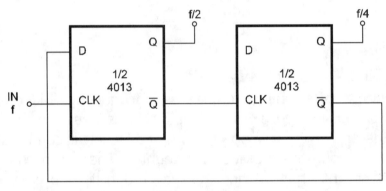

(Figure 3.68)

Cascading the 4027

A divide-by-4 or two-bit binary counter can also be built using two flip-flops of the 4027 type, as shown in Figure 3.69. It is also possible to extend the division by 8, 16, 32, 64, etc., using more flip-flops like those

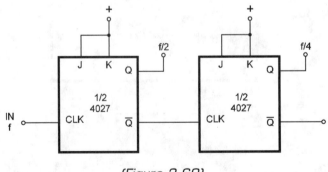

(Figure 3.69)

Basic Blocks Using CMOS ICs

found in the 4027 (Dual J-K Master/Slave Flip-Flop with Set and Reset). The maximum counting frequency is 12.5 MHz when the circuit is powered with 10 V.

Cascading n 4013

A binary counter until 2^n or a divide-by-the-same value can be built by cascading n 4013 flop-flops, as shown in Figure 3.70. Because each 4013 has two flip-flops, you will need n/2 flip-flops for a division by n even and n/2+1 for a division by n odd. The maximum input frequency depends on the power-supply voltage and is 12.5 MHz for 10 V.

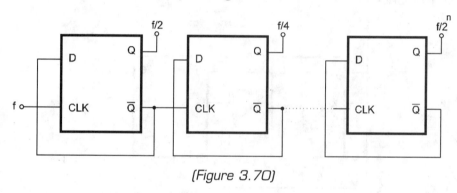

(Figure 3.70)

Cascading n 4027

Figure 3.71 shows how n 4027 can be cascaded to form a divide-by-2^n. The maximum input frequency depends on the power-supply voltage and is 12.5 MHz for 10 V.

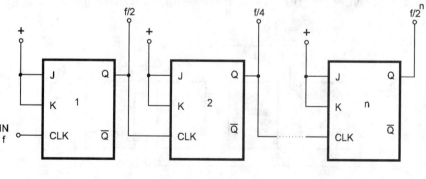

(Figure 3.71)

Shift Register Using the 4027

Data applied to the input of this circuit is shifted one cell to right following the clock pulse. The circuit is shown in Figure 3.72. A shift register is formed by D-type flip-flops cascaded, as shown in the figure. Each stage stores one bit of data. Data is applied to the input and shifted to right with the clock pulse. After a number of pulses equal to the number of stages, data applied in the input appears in the output. Data can also be available in the output of each stage. This kind of shift register that has a serial input for the signals and parallel outputs is also called SIPO (Serial In – Parallel Out).

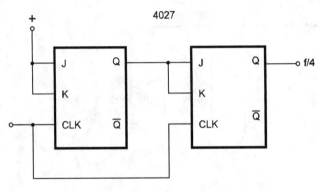

(Figure 3.72)

Shift Register Using the 4013

This is a Serial In-Parallel Out or Serial Out (SIPO or SISO) shift register using the flip-flops found in the 4013. The circuit is shown in Figure 3.73. At each cycle of the clock, data applied to the input is shifted one

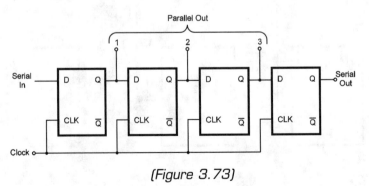

(Figure 3.73)

cell to right. Shift registers can be used as memory. The number of bits stored depends on the number of flip-flops. The maximum input frequency is 12.5 MHz when the circuit is powered from a 10 V supply.

Complete Shift Register Using the 4027

Figure 3.74 shows how n flip-flops can be associated to perform a shift register. Data is shifted one cell with each pulse of the clock. This is a SIPO or SISO. This circuit can be used to store a n-bit word. The block in the input is any inverter, such as the 4049. The maximum input frequency depends on the power-supply voltage and is 12.5 MHz with 10 V. For more information, see Section 2.

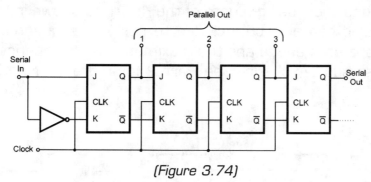

(Figure 3.74)

Divide-by-3 Counter

The circuit in Figure 3.75 shows how two flip-flops of a 4027 can be connected to perform a synchronous counter or divide-by-3. If a square

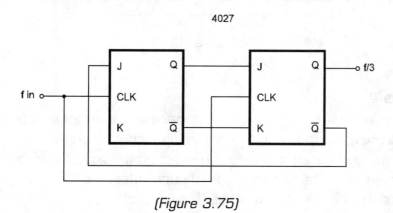

(Figure 3.75)

wave is applied to the input of this circuit, the output will be a square wave with 1/3 of the input frequency. From the waveforms shown in the figure, the reader can see that the output is not a perfect square wave (50 percent duty cycle). These states can be decoded by a 4001 as shown in the same figure. The maximum input frequency depends on the power-supply voltage and is 12.5 MHz for 10 V.

Divide-by-5 Counter

Figure 3.76 shows how the flip-flops of a 4027 can be connected to perform a divide-by-5 counter (odd division). The output isn't a perfect square wave since the duty cycle isn't 50 percent. The circuit needs some additional logic using a gate of the 4001 as shown in the figure. Additional logic can also be included to perform sequential outputs. Three flip-flops are needed and the maximum input frequency depends on the power-supply voltage. See Part 2 for more information about the 4027.

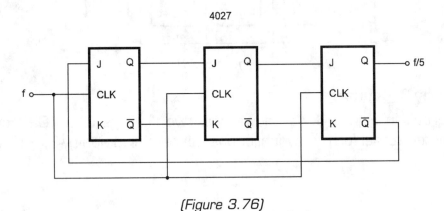

(Figure 3.76)

One-and-One

The circuit shown by Figure 3.77 generates one clock interval when a positive transition command pulse is applied to its input. The circuit is intended for applications where the synchronization of a circuit is important. The waveforms shown in the figure offer a better idea about the function of this circuit. The first flip-flop is used to cancel the difference of

Basic Blocks Using CMOS ICs

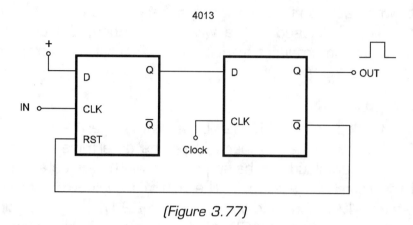

(Figure 3.77)

time between the arrival of a signal and the edge of the system clock. The second flip-flop is used to generate a pulse clock with a constant duration that will also reset the first stage.

Sequencer Using the 4013

Pulses in sequence can be generated from a clock input using the circuit shown in Figure 3.78. Outputs 1, 2, and 3 go to the high logic level in sequence as commanded by the input pulses. The number of flip-flops of the 4013 used in a design can be extended to an infinite number of channels. If you need up to 10 channels, it would be better use the 4017. This kind of circuit is also called the *bucket brigade*. It is suitable for coded switches or alarms where

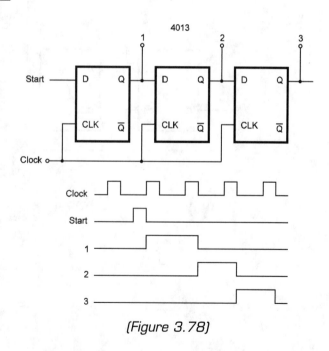

(Figure 3.78)

the next switch acts on the circuit only if the previous circuit is activated. If a set of keys is closed in the wrong sequence, the circuit will not operate. The maximum input frequency is 12.5 MHz when the circuit is powered from a 10 V supply.

Stair Generator (I)

The D-type flip-flops of a 4013 can be used to generate a stair signal as Figure 3.79 shows. The number of flip-flops determines the number of stair steps. The amplitude of the steps — or width — is determined by the value of the reference resistors. The minimum value recommended for the resistor is 47 k ohm and the maximum 22 M ohm. This circuit is a digital-to-analog converter (DAC) with a serial input. The number of pulses applied to the input will determine the voltage in the output. The maximum input frequency is 12.5 V with a 10 V supply. The maximum output frequency will be the input frequency divided by the number of steps.

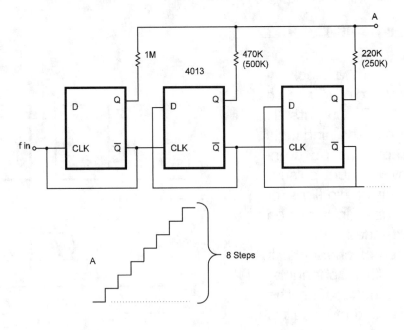

(Figure 3.79)

Basic Blocks Using CMOS ICs

Stair Generator (II)

This stair generator uses an R/2R network to determine the amplitude of the steps. This circuit is also a Digital-to-Analog converter (DAC). The number of steps is the number of flip-flops used in the configuration. The values of the resistor can be changed depending on the application, but it is not recommended to use values lower than those indicated in the diagram of Figure 3.80. The values for the resistors in the diagram can be increased depending on the application. The precision in the conversion depends on the tolerance of the resistors in the R/2R network.

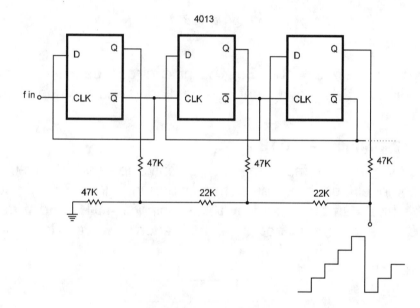

(Figure 3.80)

Divide-by-2 Counter – 4018

One flip-flop of the 4018 can be used as a frequency divider or counter, as shown in Figure 3.81. Frequency dividers or counters are very useful in many applications involving instrumentation, clocks, and chronometers. Remember that a divide-by device can perform with successive divisions of the number broken down in integer factors. For instance, a divide-by-12 can be done by successive divisions by 2, 2, and

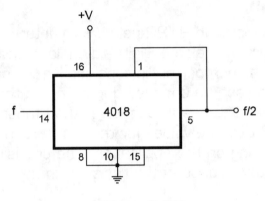

(Figure 3.81)

3 because 2 x 2 x 3 = 12. So, the next circuit can be combined to perform division by integers that can be divided into a group of numbers between 2 and 10.

Divide-by-3 Counter – 4018

To divide by an odd number means not only the need for flip-flops, but some additional logic too. The configuration in Figure 3.82 shows how a digital signal can be divided by three using a 4018 and a gate of a 4081 IC. The output is not a perfect square wave. Combining this

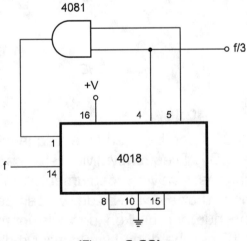

(Figure 3.82)

Basic Blocks Using CMOS ICs

circuit with the previous one, a division by six can be done. Dividers are useful circuits in the design of clocks, instruments, and many other applications. The maximum input frequency depends on the power-supply voltage. Look in Section 2 for more information about the characteristics of the 4018. The maximum input frequency is 9 MHz when the circuit is powered with 10 V.

Divide-by-4 Counter – 4018

The circuit shown in Figure 3.83 is designed to perform divisions by four of digital signals. Because the division is for an even number only, two flip-flops are needed. The output is not a perfect square wave. Additional logic may be used if a 50 percent duty-cycle signal is needed. Then maximum input frequency depends on the power-supply voltage and is 12.5 MHz with 10 V.

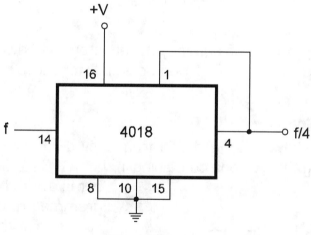

[Figure 3.83]

Divide-by-5 Counter – 4018

The division for an odd number needs the aid of some external logic in addition to the use of the 4018. Figure 3.84 shows how to make a digital frequency divider for this number using the 4018 and one gate of the 4081. The output is not a square wave just as in the previous

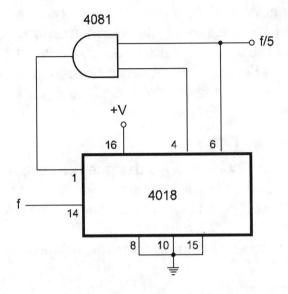

(Figure 3.84)

circuits. The maximum frequency depends on the power-supply voltage and is about 9 MHz for 10 V. See Section 2 for more information about the 4018.

Divide-by-6 Counter – 4018

Figure 3.85 shows how a digital frequency divider or counter can be created, using the flip-flops containing in the 4018 IC. Because the number is even, the counter is programmed to do the division by six without the need of additional logic. In this application, only three of the internal stages of the 4018 are used. See Section 2 for more information about the 4018 and the maximum operation frequency.

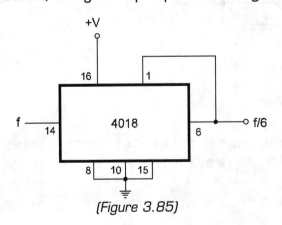

(Figure 3.85)

Basic Blocks Using CMOS ICs

Divide-by-7 Counter – 4018

Using a 4018 IC and one gate of the 4081, it is possible to divide the frequency of digital signals by seven. The circuit shown in Figure 3.86 is the configuration for this task. Because seven is an odd number, some external logic is needed. In this case, one gate of a 4081 is used. The maximum input frequency when using a power supply of 10 V is 9 MHz.

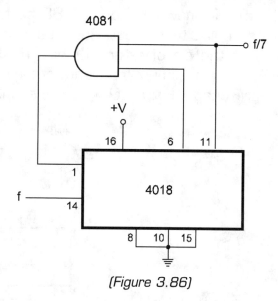

(Figure 3.86)

Divide-by-8 Counter – 4018

The division by eight using the 4018 is made as shown in Figure 3.87. It isn't necessary to use additional logic because eight is an even number. The maximum input frequency with a 10 V power supply is 9 MHz. See Section 2 for more information about the 4018.

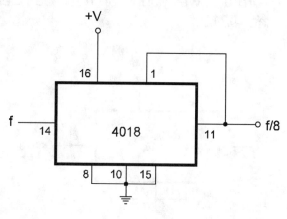

(Figure 3.87)

CMOS Sourcebook

Divide-by-9 Counter – 4018

The circuit shown in Figure 3.88 performs division by nine on a digital signal applied to its input. Because nine is an odd number, it is necessary to have the aid of external logic. This is done by using one gate of a 4081. The maximum frequency of the input signal depends on the power-supply voltage. For a 10 V power supply, the limit is about 9 MHz.

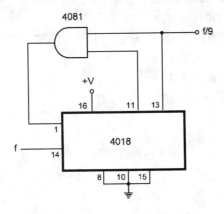

(Figure 3.88)

Divide-by-10 Counter – 4018

One 4018 is used to divide the frequency of a digital signal by 10 as shown in Figure 3.89. The maximum input frequency is 9 MHz when the circuit is powered from a power supply of 10 V. See Section 2 for more information about the 4018.

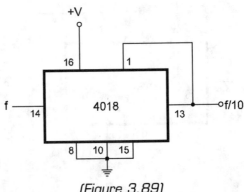

(Figure 3.89)

Basic Blocks Using CMOS ICs

Decade Counter Using the 4017

The basic configuration for a 4017 when used as a decade divider with one-of-10 decoded outputs is shown in Figure 3.90. For each pulse applied to the clock input, one output goes to the high logic level and, at the same time as the previous output, the pulse at the high logic level returns to low logic level. The circuit is reset when "1" is applied to the RST input. In normal operation, this input remains at "0". The count advances with the positive transition of the clock. The maximum input frequency depends on the power-supply voltage and is 5 MHz with a power supply of 10 V. The circuit advances in the count with the positive transition of the input signal.

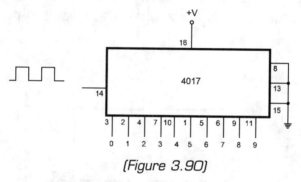

(Figure 3.90)

N-Counter Using the 4017

Counting up to numbers lower than 10 (between two and nine) can be done by connecting the n+1 output to the RST, as shown in Figure 3.91.

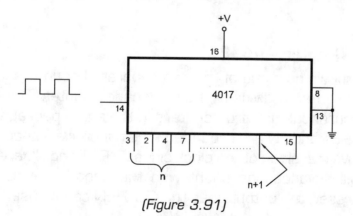

(Figure 3.91)

CMOS Sourcebook

When the counting reaches n in the next positive transition of the clock, the circuit is reset, returning to the state in which output "0" goes to the high logic level and all others remain at "0". The maximum input frequency is 5 MHz when the circuit is powered from a 10 V source.

Power-On Reset for the 4017

When the power is turned on, the output that goes to the high logic level is not necessarily the output "0". To guarantee that when the power is turned on the counting begins from the output "0" going to the "1" state and all others remain at "0," the configuration shown in Figure 3.92 is recommended. The resistor may assume values between 10 k ohm and 100 k ohm and capacitor values between 0.01 uF and 0.47 uF. The maximum input frequency is 5 MHz when the circuit is powered from a 10 V source.

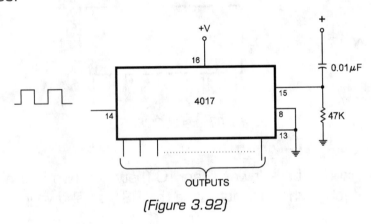

(Figure 3.92)

Divide-by-10 Counter – 4518

Figure 3.93 shows how one of the two separated counters found inside the 4518 is used as a divide-by-10 or decade counter. The 4518 contains two synchronous upward counters in the same package. In normal applications, RST and EN are grounded. The counter advances one count with the positive transition of the clock signal. If RST and CL are grounded, the circuit will advance one count with the negative transition of the clock. When reset, the output goes to 0000. EN can be used to transfer

Basic Blocks Using CMOS ICs

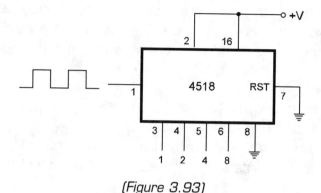

(Figure 3.93)

the ninth count to the next counter if other units are cascaded. The maximum clock frequency depends on the power-supply voltage. For a 10 V power supply, the limit is 6 MHz. See Section 2 for more information about the 4518.

Divide-by-16 Counter – 4520

The normal use of a 4520 as a divide-by-16 counter is shown in Figure 3.94. In the normal operation, RST is grounded and connected to the positive supply. The count advances with the positive transition of the clock signal. The outputs goes to 0000 if RST is made "1". The output EN presents a "1" when the count reaches 15. This can be used to

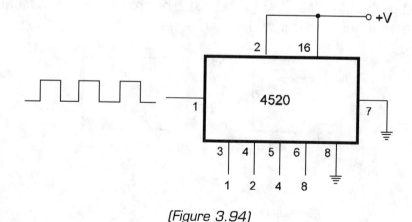

(Figure 3.94)

CMOS Sourcebook

cascade other devices. The maximum input frequency depends on the power-supply voltage and is about 6 MHz when the circuit is powered from a 10 V supply. See Section 2 for more information about the 4520.

Cascading the 4518

The 4518 can be cascaded to count to values higher than 10. The output EN of this device can be used for this task, as shown in Figure 3.95. Two devices can be used to count up to 99. The maximum input frequency is the same as in the applications where these devices stand alone.

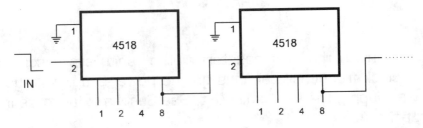

(Figure 3.95)

Cascading the 4520

For counting to values higher than 16, many 4520 ICs can be cascaded, as shown in Figure 3.96. The EN output that goes to the high logic level when the count reaches 15 is used in this case. Two 4520s can be cascaded to count up to 256 (16 x 16). The maximum input frequency is the same as when the device is used alone.

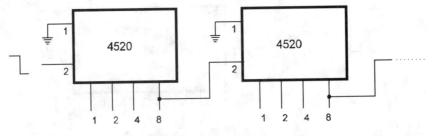

(Figure 3.96)

Basic Blocks Using CMOS ICs

Driving a 7-Segment Display – 4511

The 4511 is a 7-segment latch driver providing up to 25 mA in each output. The typical application using the 4518 as a counter is shown in Figure 3.97. Typical values for the current limiting resistor are given in the diagram. If the stored input is put to "1", the last value is held in the display. When the stored input is "0", the decoder follows the count. Cascading two blocks in this manner, we can build a counter up to 99. The display is any type with common cathode. The typical value for the current limiting resistor is 220 ohm.

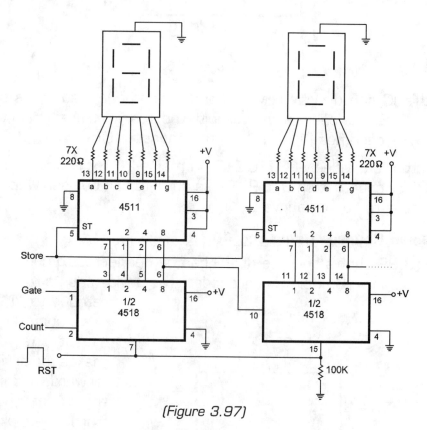

(Figure 3.97)

Driving a 7-Segment Display – 4026

The 4026 CMOS IC not only contains the complete decade counter, but also the 7-segment decoder with current limiting resources. This means

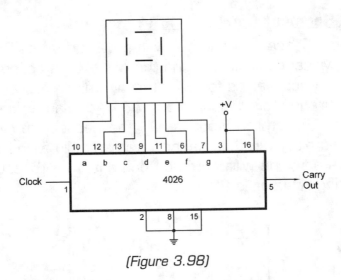

(Figure 3.98)

that this IC can directly drive a common-cathode display, as shown in Figure 3.98. The count advances with the positive transition of the clock input. RST becomes "1" to return the count to "0".

The output of divide-by-10 (10 OUT) remains "1" for counts from 0 to 4 and "0" for counts from 5 to 9. These outputs can be used when cascading devices to counts above 9.

Up/Down Counter Using the 4029

Figure 3.99 shows how the 4029 can be connected to count up to 10 and also up to 16. If pin 9 is grounded, the device counts by 10s; and if retuned to "1", the device counts by 16. Pin 10 at the "1" level makes the device count forward, and if it is "0", the count goes backward. The maximum input frequency is 7.4 MHz when the circuit is powered from a power supply of 10 V.

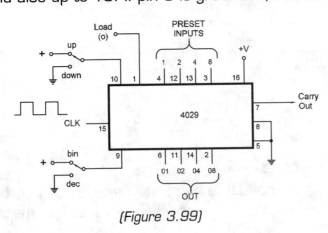

(Figure 3.99)

Basic Blocks Using CMOS ICs

Cascading the 4029

Figure 3.100 shows how many 4029s can be cascaded. The input CI of the first circuit is grounded. The output CO of this circuit is connected to the input CI of the next. This circuit is parallel clocking, but it is also possible to cascade the circuits for ripple clocking. In this case, all the inputs CI are grounded and the output CO of each block is connected to the CLK input of the next. The maximum operation frequency is 7.4 MHz when the circuit has a power supply of 10 V.

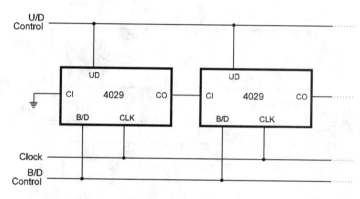

(Figure 3.100)

Using the 4510 Up/Down Counter

Figure 3.101 shows how the 4510, a divide-by-10, up/down counter with three-state outputs, is used. The count advances with the positive transition of the clock if the up/down input is "1". If the up/down input is "0", the device counts backward. The maximum count frequency is 7.6 MHz when the circuit has a power supply of 10 V. See Section 2 for more information about this device.

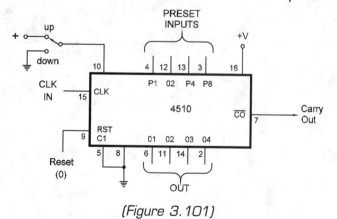

(Figure 3.101)

Cascading the 4510

Figure 3.102 shows how the 4510 can be cascaded for counts higher than 10. The circuit shows the cascading process for parallel counting. The devices can also be cascaded for ripple counting. The maximum input frequency is 7.6 MHz when the circuit has a power supply of 10 V. See Section 2 for more information about the device.

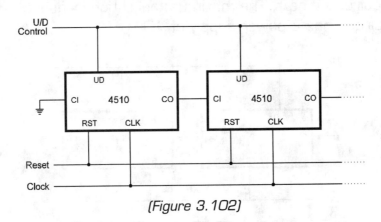

(Figure 3.102)

14-Stage Divide-by-16,384 Counter – 4020

Frequency divisions up to 16,384 can be performed with the basic configuration of a 4020, as shown in Figure 3.103. The outputs can be combined with logic gates to give intermediate values and/or high

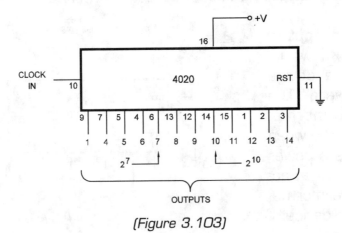

(Figure 3.103)

Basic Blocks Using CMOS ICs

values. For instance, an AND gate connected to the 2^{13} and the 2^{14} does division by 24,576. The count advances one count with the positive transition of the clock. Notice that there are no outputs for the second and third stages. The counters are reset to zero by a "1" applied to the RST input. The maximum input frequency is 10 MHz when the circuit has a power supply of 10 V.

12-Stage Divide-by-4096 Counter – 4040

Division of a digital signal up to 4096 can be done using the 4040 IC, as shown in Figure 3.104. The count advances one count with the positive transition of the clock. Counts higher than 4096 can be done by applying the signals of more than one output to appropriate logic. To reset the circuit (all outputs to 0), put RST to "1". The maximum input frequency is 8 MHz when the circuit is powered with a power supply of 10 V. See Section 2 for more information about this device.

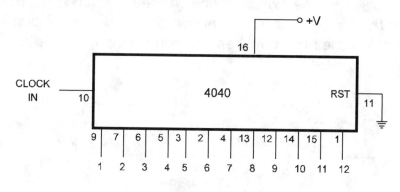

(Figure 3.104)

14-Stage Divide-by 16,384 Counter with Oscillator – 4060

This counter performs the same function as the 4020, but includes an internal oscillator. The basic use for this device is shown in Figure 3.105. There are three options for external oscillators using a crystal and RC network. If the RST input is "1", all the outputs go to "0". There are no outputs for the stages 2, 4, 8, and 2048. The maximum input frequency is 8 MHz when the circuit is powered from a power supply of 10 V.

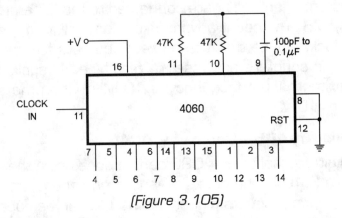

[Figure 3.105]

Analog/Digital Switch Using the 4016/4066

The 4066 is better than the 4016 because the 4016 has a lower internal resistance when the switches are on (see Section 2 for more information). In the basic application, the circuit is used as shown in Figure 3.106. The switches are on when the control input is "1". In the digital

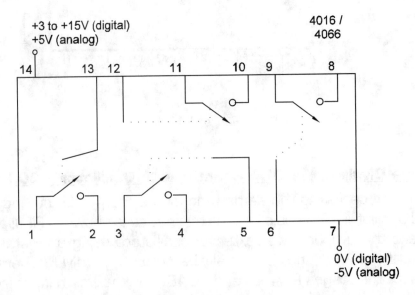

[Figure 3.106]

Basic Blocks Using CMOS ICs

mode, pin 7 is grounded. When operating with analog signals, pin 14 will be connected to a +5 V source and pin 7 to a –5 V source. The maximum amplitude of analog signals applied to the input may not exceed the power-supply voltage. The maximum control frequency is 8 MHz when the circuit is powered from a power supply of 10 V. See Section 2 for more information about the device.

Latch Using the 4066

Figure 3.107 shows a simple way to use the 4066 or 4016 as a latched switch. The output can drive stages, as many of those suggested in Part 1 show. The circuit can also be used to directly drive CMOS inputs. In the latched mode, the switch remains on after the control pulse is gone. See circuit 104 (Figure 3.104) for operation with both digital and analog signals. The maximum current through the switch may not exceed the dissipation capabilities of the device. See Section 2 for more information about the 4066 and 4016.

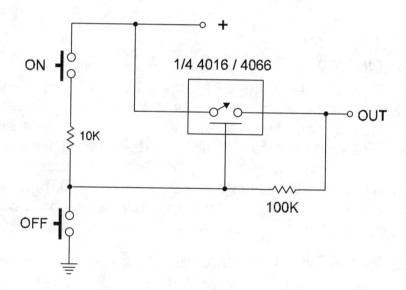

(Figure 3.107)

CMOS Sourcebook

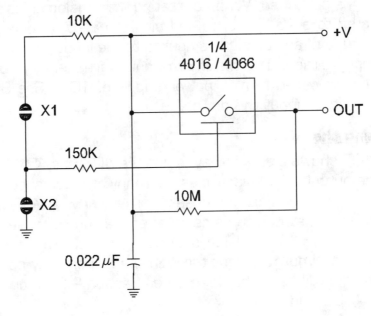

(Figure 3.108)

Touch Switch – 4066

Figure 3.108 shows how the 4066 or 4016 can be used as a touch switch. Two metal plates separated by 1 mm form the sensor. When both are touched at the same time, the two plates act as a switch and turn to a low-resistance state. See data for the 4066 in Section 2 for the currents to be controlled by this circuit. The circuit can operate with digital or analog signals. See circuit 105 (Figure 3.105) to see how to power the circuit in the two-operation mode. When working with analog signals, it is important to avoid noise in the control input.

Analog-to-Digital Converter with the 4066

A four-bit analog-to-digital converter (A/D or ADC) can be built using one 4066. An 8-bit converter can be designed using two chips. The input network using the resistor determines the steps in the conversion. Figure 3.109 shows the diagram for this basic converter. Typical values for

Basic Blocks Using CMOS ICs

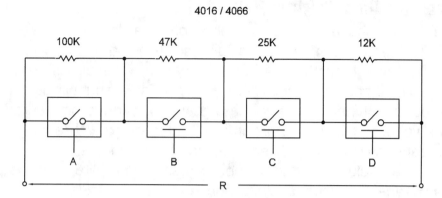

(Figure 3.109)

the resistor are given in the circuit. The resistor connected to the circuit will depend on which switch is turned on. This circuit can be used as part of a digital volume control or a digital power supply. The maximum input frequency for this circuit is 40 MHz.

Digital Variable Capacitor

The values of the capacitance between output A and ground can be programmed digitally in 16 steps as determined by the values of the four capacitors in the network. If the capacitance must not start from zero, another capacitor (C5) may be used. The circuit shown in Figure 3.110 can be used to perform as a digital controlled oscillator. In the same figure, it suggests another application for this circuit — controlling the gain of an operational amplifier.

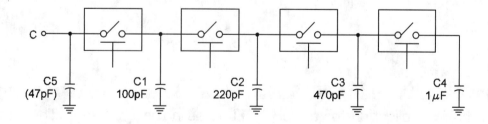

(Figure 3.110)

One-of-Eight Data Selector – 4051

Figure 3.111 shows the basic application of a data selector using the 4051. The same configuration is valid for other data selectors, such as the 4052 (one-of-four) and 4053 (one-of-two). The digital number programmed in the address input determines which output will be connected to the output. When working with digital data, Vdd is connected to the positive voltage and Vss to ground. When working with analog signals, Vdd is connected to a +5 V source, Vss to ground, and Vee to a –5 V source. The analog signal must not exceed the amplitude of the power-supply voltage. The frequency response of each channel when on is 40 MHz when the circuit is powered from a supply of 10V.

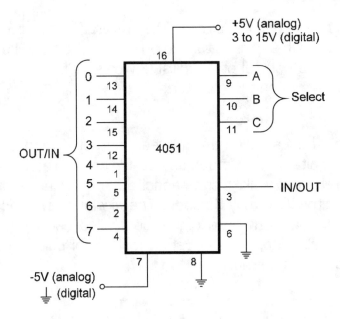

(Figure 3.111)

Tone Filter Using the 4066 PLL

The basic application for a 4066 PLL is as a tone detector. The circuit can be selected to function in two modes — with high or low selectivity. Suitable applications are in the range between a low hertz and more

Basic Blocks Using CMOS ICs

than 2 MHz. The circuit of a tone detector using the 4066 is shown in Figure 3.112. This circuit can recognize tones in a narrow frequency band. When a tone is recognized, the circuit locks and a control pulse is available in the outputs. In this case, the circuit has two outputs with symmetrical signals when locked. This symmetry can be detected by a logic gate sourced to the output after a filter (command pulse). The loop filter, which determines the characteristics of the PLL, is formed by R1, R2, and C1. C1 and R1 determine the setting time; the ratio of R1 to R2 determines the damping factor. R3 and C2 determine the maximum VCO frequency, and the minimum is given by R1, R2, and C2. The values shown in the diagram are typical for a detection of a 4 kHz tone.

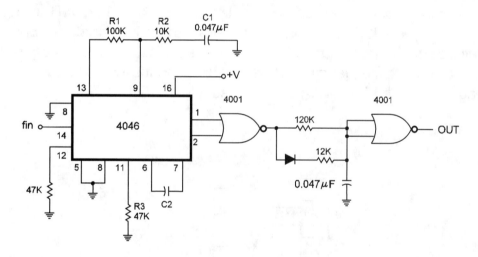

(Figure 3.112)

Complete Applications

The following circuits can stand alone as complete devices. The circuits are basic and many components can be changed to alter the performance of the circuit.

Digital Amplifier

When more current from a digital output is needed than a single output could source or drain, a digital amplifier can be used. In this case, the gates as inverters can be connected in a parallel manner to increase the devices capabilities. The circuit shown in Figure 3.113 is an example of digital amplifier and buffer that uses four gates of a 4093 IC or any other IC that can be connected as an inverter (4001, 4011, 4046, etc.). Six inverters of a 40106 can also be used in the same configuration. Another use for digital buffers is providing isolation between stages. The maximum input frequency for this circuit depends on the power-supply voltage and is 10 MHz with a power supply of 10 V.

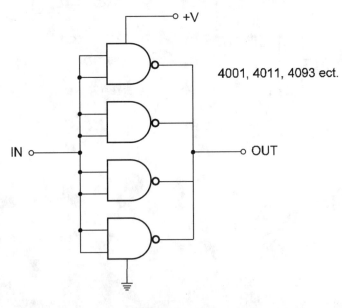

(Figure 3.113)

Basic Blocks Using CMOS ICs

Bridge Digital Amplifier

The circuit shown in Figure 3.114 shows how four inverters (four NAND, four NOR, or four Inverters) can be connected to form a digital amplifier. When the input is "1", the current flows in one direction by the load, and when the input is "0", the current flows in the opposite direction. This circuit is ideal to drive LEDs or other loads that have polarity. Two opposite paralleled LEDs can be driven by this stage. The LED that will remain on dependent on the logic level applied to the input.

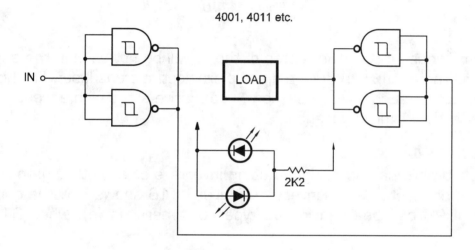

(Figure 3.114)

Two-Tone Siren

The gates of a 4093 can be used form a two-tone siren, as shown in the configuration of Figure 3.115. The tone is given by C2/R2 and the modulation by C1/R1. R1 and R2 can be experimented with in a large range of values. If you would like a frequency adjustment for the oscillators, use trimmer potentiometers to replace the resistors. A buffer is

CMOS Sourcebook

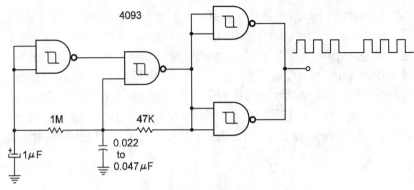

(Figure 3.115)

recommended if the other gates of the IC will be not used in the application. The circuit can drive a small piezoelectric transducer. For high-power loads, such as loudspeakers, use a drive stage with a transistor (see Section 1 for suggestions).

Timers with Inverters

Simple timers can be designed using inverters or any other gate that can be connected as an inverter. Figure 3.116 shows how the gates of a 4093 can be used in two types of timers. In (a), when S1 is

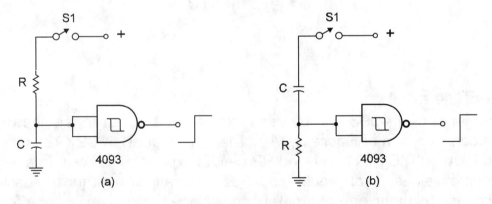

(Figure 3.116)

Basic Blocks Using CMOS ICs

closed, the output of the IC goes to "1" during a time interval determined by R1 and C1. After this interval, the output goes to "0". In (b), the output goes to "0" when the switch is closed, and after the time interval determined by R1/C1, goes to "1". R1 can be in the range between 10 k ohm and 22 M ohm. The minimum value for C1 is 100 pF. These timers can provide time delays up to 30 minutes depending on the value of the components used.

Alarm

When the sensor X1 is opened, the relay is energized. When waiting, the current drained is very low. Several sensors can be connected in series to protect different places. The sensor can be a reed switch or a thin wire that is interrupted when a door or a window is opened. The circuit is shown in Figure 3.117. The relay is a sensitive type with a coil depending on the power-supply voltage and current should not exceed 50 mA. Another configuration derived from this is obtained by connecting the output of the second 4093 to the control input of the siren pictured in Figure 3.113. This configuration will eliminate the need for a relay.

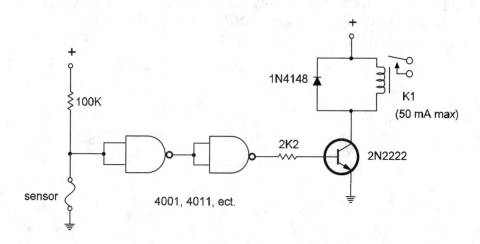

(Figure 3.117)

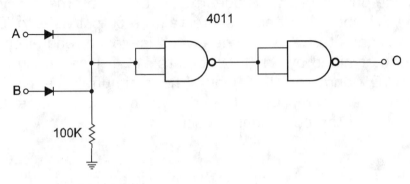

(Figure 3.118)

OR Gate Using NAND

An OR gate can be improvised using NAND gates of a 4001, as shown in Figure 3.118. The diodes are general-purpose silicon types, such as the 1N4148 or 1N914. More inputs can be added with additional diodes. OR Schmitt gates can be created with the use of a 4093.

AND Gate Using NOR

Three of the four gates found in a 4001 can be used to perform as an AND gate. This improvisation can be useful if you don't have handy an AND gate for a project or have a 4001 in the circuit with three of its gates unused. The circuit is shown in Figure 3.119.

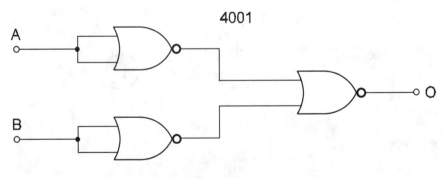

(Figure 3.119)

Basic Blocks Using CMOS ICs

NOR Gate Using NAND

A NOR gate can be improvised using one NAND gate of a 4011 as shown in Figure 3.120. The diodes are general-purpose types, such as the 1N914 or 1N4148. Inverters or even Schmitt triggers, such as the 4093 or 40106, can be used in this configuration. If more inputs are needed, simply add more diodes to the input.

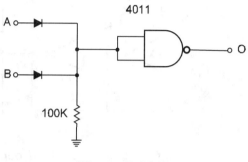

(Figure 3.120)

Cricket

This circuit uses three oscillators to produce a sound similar to a cricket chirping. Two oscillators act as modulator for the third, resulting in a tone that is produced by a small piezoelectric transducer. The tone is a "chirp" like a cricket makes. The circuit shown in Figure 3.121 has a very low current drain and can be powered from a 9 V battery. The sound is even more realistic if low-power transducers are used. For a small loudspeaker, use a drive stage with a transistor (see Section 1 for more information).

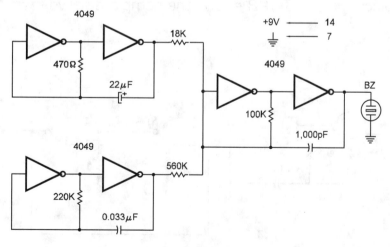

(Figure 3.121)

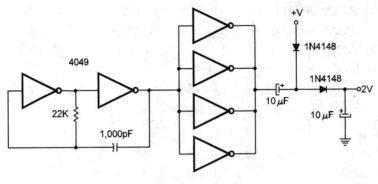

(Figure 3.122)

Voltage Doubler

The simple circuit doubler shown in Figure 3.122 can source a voltage two times the one applied to the input. The output current is very low, indicating that the device is suitable only for bias purposes. The diodes are general-purpose types, such as the 1N914 or 1N4148.

Servo Test

Servo motors operate from the changes in the pulse width in a range between 1 and 2 ms. The separation between pulses is 20 ms. This circuit is used to produce variable pulses in the range needed to test these devices. Figure 3.123 shows the complete servo test. The circuit

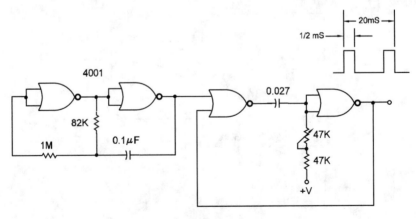

(Figure 3.123)

Basic Blocks Using CMOS ICs

is formed by an oscillator that determined the central frequency of 50 Hz and a monostable that produces pulses with variable lengths. The pulse width is adjusted by the 47 k ohm potentiometer within the range that needs to be tested.

Who is the Fastest?

Only one LED is turned on when S1 and S2 are pressed. The LED that will be on is just the correspondent to the switch that is activated first. This circuit can be used in contests or competitions. Sensors, such as reed switches, can be used to replace the switches. The RST switch is used to reset the circuit after a contest. The circuit is shown in Figure 3.126. If lamps are to be used in the application, a drive stage can be used. See Section 1 for suggestions. The basic configuration can be extended to include more than two switches.

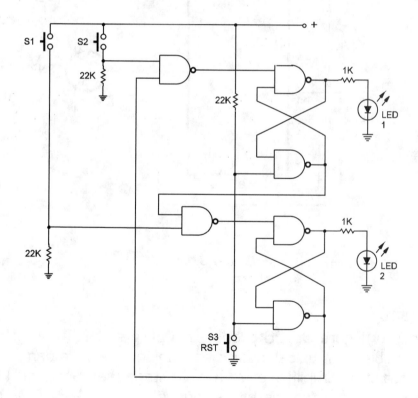

(Figure 3.124)

Bistable Relay – 4013

A negative pulse applied to the input of the circuit shown in Figure 3.125 turns on the relay. The next pulse will turn off the relay. The bistable action of the block is obtained thanks to the use of one of the two flip-flops of a 4013. The relay can be any type, depending on the power-supply voltage and a current up to 50 mA. In this example, the inputs of the unused flip-flop are grounded. The pinout of the 4013 can use the other flip-flop of the 4013 for other applications, including a second bistable.

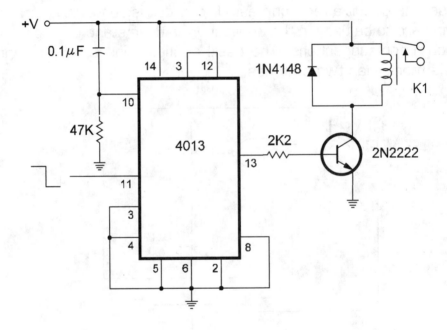

(Figure 3.125)

Tuning Fork

The circuit of Figure 3.126 produces a pure square tone of 440 Hz that is suitable for musical instrument tuning. The tone can be applied to the input of an amplifier or to a drive stage (see Section 1) using a BD135 or another transistor as loud a loudspeaker. The trimmer

Basic Blocks Using CMOS ICs

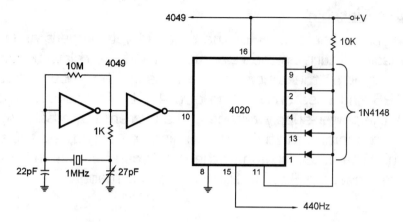

(Figure 3.126)

capacitor adjusts the point for a correct start for the oscillator when the power is turned on. The diodes are general-purpose types, such as the 1N914 or 1N4148.

Frequency Standard

The circuit shown in Figure 3.127 generates frequencies of 50 Hz, 100 Hz, and 200 Hz. The circuit uses an XTAL of 3278 kHz and two CMOS ICs: a 4060 to produce a 200 Hz signal and a 4013 (two flip-flops) to divide this frequency by 2 and 4. Using two stages for division by 5 and 10, you can obtain 1 Hz pulses from the 50 Hz signal for a clock or chronometer.

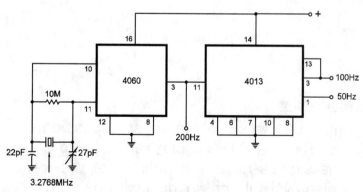

(Figure 3.127)

CMOS Sourcebook

Touch/Light Bistable

A flash of light on one sensor or the touch of your fingers on the other type of sensor will turn on the relay. To turn off the relay, simply repeat the flash of light or the touch on the sensor. The circuit is shown in Figure 3.128 and uses one flip-flop of a 4013. Any relay with coil, depending on the power-supply voltage and a current up to 50 mA, can be used. The potentiometer in the version using an LDR adjusts the sensivity of the circuit. The other flip-flop of the 4013 can be used in other applications. See pinouts in Section 2 for more information.

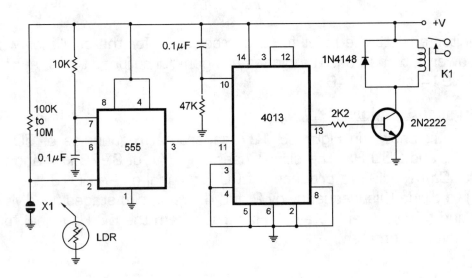

(Figure 3.128)

Two-Phase Generator

This circuit generates two signals in phase opposition. The frequency is determined by R1 and C1 and can reach up to 1 MHz in typical applications. The signals have one quarter of the frequency of the oscillator. It is also available as a signal with half of the frequency of

Basic Blocks Using CMOS ICs

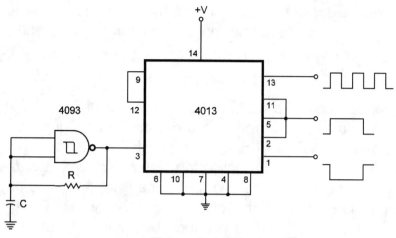

(Figure 3.129)

the oscillator. The circuit is shown in Figure 3.129. The oscillator used in the 4093 can be replaced by the oscillators used in circuit 124 or 125 (Figures 3.124 and 3.125).

Frequency Divider Using the 4017

Decade counters, such as the 4017, can be cascaded as sourcing signals with the frequency divided by 10, 100, 1000, etc. The circuit is shown in Figure 3.130 and is an example of this application. It operates

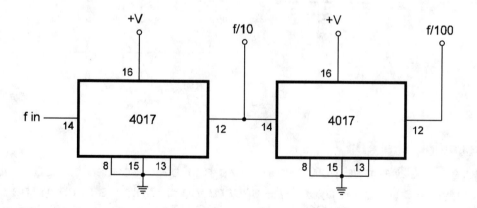

(Figure 3.130)

only with digital signals. Remember that the output is not a square wave. If you need a square wave (50 percent of duty cycle), additional logic sources may be used.

Sequencer

The basic application of a 4017 as a sequencer (10 steps) is shown in Figure 3.131. In this same part, you can find a configuration using the 4017 to count up to less than 10. The 555 generates the pulses to drive the 4017. In the basic application, you can drive LEDs, transistors, Relays, SCRs, and many other devices. See Part 1 for the drive stages that can be connected to the outputs of this circuit. Other oscillator configurations can be used to produce the pulses of the sequence. If sensors or switches generate the pulses, they must be debounced. The use of a 555 monostable is recommended for this task.

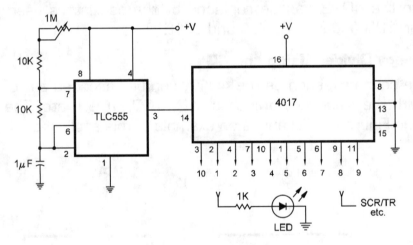

(Figure 3.131)

Cascading the 4017

Figure 3.132 shows how two 4017s can be cascaded to count above 10. The circuits are driven by a square input, which is seen in the oscillator shown in circuit 129 (Figure 3.129). The same loads as shown in the previous projects can be connected to the output of this circuit.

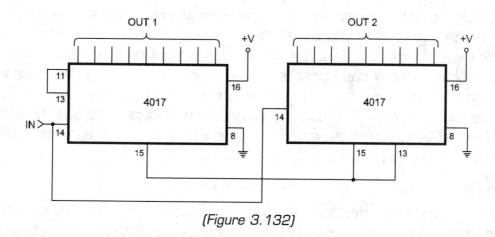

(Figure 3.132)

Digital Mixer

A stereo digital mixer using the 4053 is built, as shown in Figure 3.133. The IC must be powered from a symmetrical power supply (+5 V to +15 V and –5 V to –15 V). If working with analog signals having amplitudes up to 15 Vpp, Vss = 0 V, Vdd = 15 V, and Vee = -15 V. The inputs are multiplexed to the output in a frequency determined by the oscillator

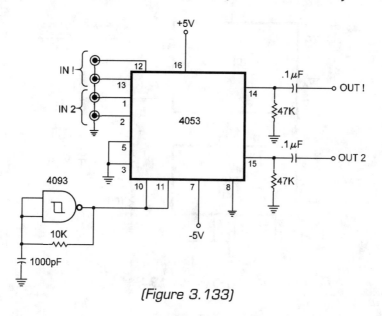

(Figure 3.133)

CMOS Sourcebook

(4093). To not generate noise in the output, clocking frequencies must be much higher than the top limit of the audio band. 100 kHz is a suitable value. In some cases, the output signal must be filtered to eliminate the high-frequency components that are produced during the switching process. Remember that this circuit operates with low-level signals from audio sources, such as preamplifiers and transducers. Because the 4051 is formed by two multiplexer/demultiplexers in one package, for stereo applications you will need only one device.

Audio Selector

S1 changes the selection of the input signals in the circuit shown in Figure 3.134. Other type of circuits, such as monostables with sensors, can generate the command. Notice that the circuit is powered from a +5 and –5 V power supply. The circuit is powered following the next procedure: Vdd = 5 V to 15 V; Vss = 0 V, and Vee = -5 to –15 V. The amplitude of the input signals must not exceed the power-supply voltage. The control input (pins 9 and 10) can be connected to other signal sources, such as the output of counters or memory.

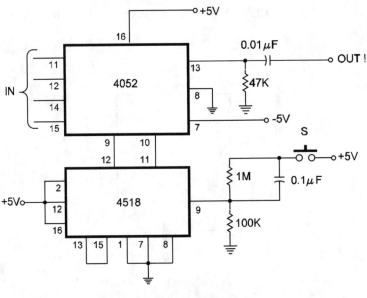

(Figure 3.134)

Index

A

A/B input 158
A/B line 158
ADC 366
addressable latch 291
AE input 158
algorithm 17
alternate action switch 339
amplifier, digital 370, 371
amplifier, operational 3
AND 8
AND gate 5, 19, 205, 213, 363, 374
analog-to-digital converter 349
astable multivibrator 225
A/S line 158
Asynchronous/Synchronous line 158

B

bilateral switch 119, 195
binary addition block 5
binary arithmetic 10
Binary Coded Decimal 10
binary counter 273, 342, 343
binary mode 150
binary numbering system 10
binary ripple counter 139
binary-coded decimal 14
bipolar transistor 62, 73, 300, 301
block, CMOS 84, 308
bus system 157
byte 13

C

carryout pulse 122
CCO 320
circuit, CMOS 3, 59, 323
circuit, monostable 323
circuit, TTL 64
circuits, logic 3
clock cycle 122, 132
clock pulse 132, 161
clocked RS Latch 5
CMOS block 84, 85, 308
CMOS circuit 3, 59, 323
CMOS gate 3, 5, 8, 66, 84
CMOS IC 3, 5, 57, 59, 60, 62, 299, 300, 310
CMOS IC output 62
CMOS input 300, 365

CMOS inverter 5, 57
CMOS logic 5, 308
CMOS logic circuits 3
CMOS monostable 323
CMOS multivibrator 331
CMOS output 58
combinational logic 5
converter, digital-to-analog 349
converter, analog-to-digital 366
counter, binary 273
counter, decade 257
counter, ripple 140
crystal oscillator 322
current-controlled oscillator 320

D

D Latch 5
D-type flip-flop 209, 328, 344, 348
D-type latch 165
DAC 349
Darlington transistor 67
decade counter 257
decade mode 150
decoder 5
DeMorgan Theorem 44
demultiplexer 5, 52, 224, 384
demultiplexing 188, 191, 292
DEMUX 52
digital amplifier 370, 371
digital frequency divider 352
Digital logic 4
digital-to-analog converter 349

diode, parallel 66
discrete transistor 70
DSP 3
dual-retriggerable monostable 330
duty cycle 311, 346

E

Edge-Triggered RS Flip-Flop 48
electromagnetic interference 63
EMI 63

F

FET 62
FET, Power 78
flip-flop 74, 113, 143, 167, 239, 242, 332, 336
flip-flop, D-type 209, 328, 344, 348
flip-flop, J-K 5
flip-flop, R-S 5, 264

G

gate, AND 5, 19, 205, 213, 363, 374
gate, CMOS 3, 5, 8, 66, 84
gate, LS TTL 185
gate, NAND 5, 7, 8, 22, 109, 111, 137, 138, 303, 306, 325, 327, 328, 336, 338, 374

Index

gate, NOR 5, 6, 25, 46, 94, 95, 97, 105, 141, 142, 202, 313, 327, 337
gate, OR 5, 23, 27, 96, 152, 199, 201, 202, 207, 374
gate, TTL 185
gated oscillator 312

H

hexadecimal 10, 16
hexadecimal numbering system 16
high-power SCR 76
hysteresis characteristic 220

I

IC, CMOS 3, 5, 57, 59, 60, 62, 299, 300, 310
IGBT 78
IGFET transistor 59
IN input 132
input, CMOS 300, 365
input, IN 132
input, LD 132
input logic level 21, 22
input, LS TTL 251
input, TTL 85, 198, 201, 208, 251
input, TTL LS 198, 208
inverter 5
inverter, CMOS 5, 57
Isolated-gate bipolar transistor 78

J

J-K flip-flop 5
JAM input 115
Johnson Counter 121

L

latch, addressable 291
Latch, D 5
Law of Double Negation 40
Laws of Absorption 35
Laws of Association 33
Laws of Commutation 32
Laws of Complementation 38
Laws of Contraposition 39
Laws of Distribution 34
Laws of Duality 42
Laws of Expansion 40
Laws of Null Class 37
Laws of Tautology 31
Laws of Universe Class 36
LD input 132
LDR 304, 319
least significant bit 162
light dependent resistor 304
logic circuits 3
logic, CMOS 5
logic, combinational 5
logic, Digital 4
logic level 94
logic, sequential 5
logic signal 94, 98, 110
LS TTL gate 185
LS TTL input 251

M

master-slave flip-flop 48
microcontroller 3
microprocessor 3
monostable 323, 327, 328, 330, 334, 377, 382
monostable circuit 323
monostable, CMOS 323
monostable, dual-retriggerable 330
monostable multivibrator 226, 276, 282, 330, 332
monostable, NOR 328
monostable, retriggerable 335
MOS transistor 5, 6, 101
MOSFET 6, 8, 72
MOSFET, power 73
most significant bit 162
multiplexer 5, 52, 108, 384
multiplexing 188, 191
multiplexing/demultiplexing 281
multivibrator 225, 226, 334
multivibrator, astable 225
multivibrator, CMOS 331
multivibrator, monostable 226, 276, 282, 330, 332
MUX 52
Mux/Demux 188, 189, 192, 281

N

N-channel 6
nalog-to-digital converter 366

NAND Gate 22
NAND gate 5, 7, 8, 22, 109, 111, 137, 138, 303, 306, 325, 327, 328, 336, 338, 374
NAND input 169
NAND Schmitt Gate 339
network, R/2R 349
NOR gate 5, 6, 25, 46, 94, 95, 97, 105, 141, 142, 202, 313, 327, 337
NOR input 172
NOR monostable 328
NPN transistor 66, 69, 77, 85

O

operational amplifier 3
optocoupler 79
optodiac 81
OR Gate 152
OR gate 5, 23, 27, 96, 199, 201, 202, 207, 374
oscillator, crystal 322
oscillator, current-controlled 320
oscillator, ring 314, 318
oscillator, three-gate ring 314
oscillator, voltage-controlled 317
output, CMOS 58
output logic level 21, 22
output, TTL 64

Index

P

P-channel 6
P/S input 158
parallel diode 66
phototransistor 81, 319
PLL 173
PNP Transistor 68
potentiometer 308, 311
potentiometer, trimmer
    308, 339
Power FET 78
Power MOSFET 73

R

R-S flip-flop 264, 337
R/2R network 349
R/S latch 170
RC network 308, 310
Rds 73
retriggerable monostable 335
ring oscillator 310, 314, 318
ripple clocking 361
ripple counter 130, 140
RS Flip-Flop 5
RS NAND Latch 5
RS NAND latch 45
RST switch 378
RTR input 177

S

Schmitt inverter 220, 303,
    304, 324
Schmitt NAND gate
    303, 306, 336
Schmitt oscillator 306
Schmitt trigger
    218, 220, 225,
    309, 375
SCR 3, 62, 73, 75, 79
SCR, high-power 76
sequential logic 5
Serial In/Parallel Out 161, 344
Serial In/Serial Out 161, 344
shift register 157
SIPO 344
SISO 344
stair generator 349
switch, bilateral 119, 195
switch, RST 378
synchronous counter 227,
    230, 233, 236, 244, 247

T

three-gate ring oscillator 314
TR input 177
transistor 3
transistor, bipolar 62, 73,
    300, 301
transistor, Darlington 67

transistor, discrete 70
transistor, IGFET 59
transistor, MOS 5, 6, 101
transistor, NPN 66, 69, 77, 85
Transistor, PNP 68
Triac 3, 77, 81
trimmer potentiometer 308, 339
TTL 54
TTL block 83
TTL circuit 64
TTL gate 83, 185
TTL input 85, 198, 201, 208, 251
TTL interface 106
TTL LS input 198, 208
TTL output 64

U

up-down counter 252

V

Vcc 59, 105
VCF 315
VCO 312, 315, 317
VCO Frequency 174
Vdd 60, 101, 105, 107
Voltage to Frequency Converter 315
Voltage-Controlled Oscillator 312, 317
Vss 60, 101, 105

EXPLORING LANS FOR THE SMALL BUSINESS & HOME OFFICE

Author: LOUIS COLUMBUS
ISBN: 0790612291 • **SAMS#:** 61229
Pages: 304 • **Category:** Computer Technology
Case qty: TBD • **Binding:** Paperback
Price: $39.95 US/$63.95CAN

About the book: Part of Sams Connectivity Series, *Exploring LANs for the Small Business and Home Office* covers everything from the fundamentals of small business and home-based LANs to choosing appropriate cabling systems. Columbus puts his knowledge of computer systems to work, helping entrepreneurs set up a system to fit their needs.

PROMPT® Pointers: Includes small business and home-office Local Area Network examples. Covers cabling issues. Discusses options for specific situations. Includes TCP/IP (Transmission Control Protocol/Internet Protocol) coverage. Coverage of protocols and layering.

Related Titles: *Administrator's Guide to E-Commerce*, by Louis Columbus, ISBN 0790611872. *Administrator's Guide to Servers*, by Louis Columbus, ISBN

Author Information: Louis Columbus has over 15 years of experience working for computer-related companies. He has published 10 books related to computers and has published numerous articles in magazines such as *Desktop Engineering, Selling NT Solutions*, and *Windows NT Solutions*. Louis resides in Orange, Calif.

To order today or locate your nearest PROMPT® Publications distributor at 1-800-428-7267 or www.samswebsite.com

Prices subject to change.

EXPLORING MICROSOFT OFFICE XP

Authors: JOHN BREEDEN & MICHAEL CHEEK
ISBN: 079061233X ● **SAMS#:** 61233
Pages: 336 ● **Category:** Computer Technology
Case qty: TBD ● **Binding:** Paperback
Price: $29.95 US/$47.95CAN
About the book: Breeden and Cheek provide an insight into the newest product from Microsoft — Office XP. Office XP is the replacement for Microsoft Office, designed to take users into the 21st century. Breeden and Cheek provide tips and tricks for the experienced office user, to help them find maximum value in this new software.

ELECTRONICS FOR THE ELECTRICIAN

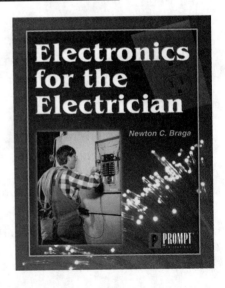

Author: NEWTON C. BRAGA
ISBN: 0790612186 ● **SAMS#:** 61218
Pages: 320 ● **Category:** Electrical Technology
Case qty: 32 ● **Binding:** Paperback
Price: $34.95 US/$55.95CAN
About the book: Author Newton Braga takes an innovative approach to helping the electrician advance his or her career. Electronics have become more and more common in the world of the electrician, and this book will help the electrician become more comfortable and proficient at tackling the new tasks required.

To order today or locate your nearest PROMPT® Publications distributor at 1-800-428-7267 or www.samswebsite.com

Prices subject to change.

APPLIED SECURITY DEVICES & CIRCUITS

Author: PAUL BENTON
ISBN: 079061247X • **SAMS#:** 61247
Pages: 280 • **Category:** Projects
Case qty: TBD • **Binding:** Paperback
Price: $34.95 US/$55.95CAN

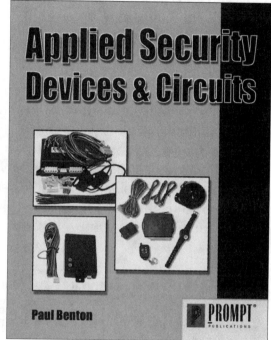

About the book: The safety and security of ourselves, our loved-ones and our property are uppermost in our minds in today's changing society. As security components have become user-friendly and affordable, more and more people are installing their own security systems. Paul Benton covers this topic in a "secure" way, applying proven electronics techniques to do-it-yourself security devices.

Prompt Pointers: Includes automobile security systems, basic alarm principles, and high-voltage protection. Outlines over 100 applied security applications. Contains over 200 illustrations.

Related Titles: *Guide to Electronic Surveillance Devices*, ISBN 0790612453, *Guide to Webcams*, ISBN 0790612208, *Applied Robotics*, ISBN 0790611848.

Author Information: Paul Benton has been involved in electronics since leaving school originally as a TV and radio technician, before becoming involved in electronic security devices and techniques in the 1980s. Under the name of Paul Brookes, his mother's maiden name, Benton has written a number of electronics-related books and articles. As a teacher and lecturer at the university level, Benton remains current with today's technologies and currently works for an international electronic company in England.

To order today or locate your nearest PROMPT® Publications distributor at 1-800-428-7267 or www.samswebsite.com

Prices subject to change.

ADMINISTRATOR'S GUIDE TO SERVERS

Author: LOUIS COLUMBUS
ISBN: 0790612305 • **SAMS#:** 61230
Pages: 304 • **Category:** Computer Technology
Case qty: TBD • **Binding:** Paperback
Price: $39.95 US/$63.95CAN

About the book: Part of Sams Connectivity Series, *Administrator's Guide to Servers* piggybacks on the success of Columbus' best-selling title *Administrator's Guide to E-Commerce*. Columbus takes a global approach to servers while providing the detail needed to utilize the correct application for your Internet setting.

PROMPT® Pointers: Compares approaches to server development. Discusses administration and management. Balance of hands-on guidance and technical information.

Related Titles: *Administrator's Guide to E-Commerce*, by Louis Columbus, ISBN 0790611872. *Exploring LANs for the Small Business and Home Office*, by Louis Columbus, ISBN 0790612291. *Computer Networking for the Small Business and Home Office*, by John Ross, ISBN 0790612216.

Author Information: Louis Columbus has over 15 years of experience working for computer related companies. He has published 10 books related to computers and has published numerous articles in magazines such as *Desktop Engineering, Selling NT Solutions*, and *Windows NT Solutions*. Louis resides in Orange, Calif.

To order today or locate your nearest PROMPT® Publications distributor at 1-800-428-7267 or www.samswebsite.com

Prices subject to change.

SEMICONDUCTOR CROSS REFERENCE BOOK, 5/E

Author: SAMS TECHNICAL PUBLISHING
ISBN: 0790611392 • **SAMS#:** 61139
Pages: 876 • **Category:** Professional Reference
Case qty: 14 • **Binding:** Paperback
Price: $39.95 US/$63.95CAN

About the book: The perfect companion for anyone involved in electronics! Sams has compiled years of information to help you make the most of your stock of semiconductors. Both paper and CD-ROM versions of this tool contain an additional 128,000 parts listings over the previous editions.

ON CD-ROM, 2E

ISBN: 0790612313 • **SAMS#:** 61231 • **Price:** $39.95 US/$63.95CAN

COMPUTER NETWORKS FOR THE SMALL BUSINESS & HOME OFFICE

Author: JOHN A. ROSS
ISBN: 0790612216 • **SAMS#:** 61221
Pages: 304 • **Category:** Computer Technology
Binding: Paperback • **Price:** $39.95 US/$63.95CAN

About the book: Small businesses, home offices, and satellite offices with unique networks of 2 or more PCs can be a challenge for any technician. This book provides information so that technicians can install, maintain and service computer networks typically used in a small business setting. Schematics, graphics and photographs will aid the "everyday" text in outlining how computer network technology operates, the differences between various network solutions, hardware applications, and more.

To order today or locate your nearest PROMPT® Publications distributor at 1-800-428-7267 or www.samswebsite.com

Prices subject to change.

ADMINISTRATOR'S GUIDE TO DATAWAREHOUSING

Author: AMITESH SINHA
ISBN: 0790612496 • **SAMS#:** 61249
Pages: 304 • **Category:** Computer Technology
Case qty: TBD • **Binding:** Paperback
Price: $39.95 US/$63.95CAN

About the book: Datawarehousing is the manipulation of the data collected by your business. This manipulation of data provides your company with the information it needs in a timely manner, in the form it desires. This complex and emerging technology is fully addressed in this book. Author Amitesh Sinha explains datawarehousing in full detail, covering everything from set-up to operation to the definition of terms.

PROMPT® Pointers: Covers On-Line Analytical Processing issues. Addresses set-up of datawarehousing systems. Is designed for the experienced IT administrator.

Related Titles: *Designing Serial SANS*, ISBN 0790612461, *How the PC Hardware Works*, ISBN 079061250X.

Author Information: Amitesh Sinha has a Masters in Business Administration and over 10 years of experience in the field of Information Technology. Sinha is currently the Director of Projects with GlobalCynex Inc. based in Virginia and has written numerous articles for computer publications.

To order today or locate your nearest PROMPT® Publications distributor at 1-800-428-7267 or www.samswebsite.com

Prices subject to change.

APPLIED ROBOTICS II

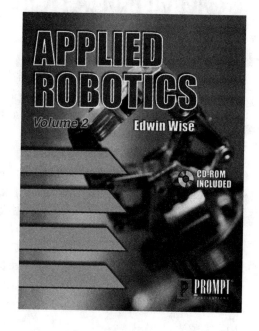

Author: EDWIN WISE
ISBN: 0790612224 ● **SAMS#:** 61222
Pages: 304 ● **Category:** Projects
Case qty: TBD ● **Binding:** Paperback
Price: $29.95 US/$47.95CAN

About the book: Edwin Wise builds upon his best-seller, *Applied Robotics* with this book targeted at more advanced hobbyists with development of a larger, more robust, and very practical mobile robot platform. Building on the foundation set in his first text, *Applied Robotics II* has projects to create a larger robot platform suitable for use in the home or outdoors, advanced sensor projects and a great exploration of A1 and control software.

Prompt Pointers: Picks up where *Applied Robotics* left off. Offers an advanced set of projects related to this very hot subject area.

Related Titles: Applied Robotics, ISBN 0790611848. Animatronics, ISBN 079061294.

Author Information: Edwin Wise is a professional software engineer with twenty years of experience. He currently works in the field of Computer Aided Manufacturing (CAM). His experience includes work on both computer games and educational software. Building robots has been a dream and passion for Edwin for years now. His current project is "Boris," a giant killer robot that can be viewed at http://www.simreal.com/Boris.

To order today or locate your nearest PROMPT® Publications distributor at 1-800-428-7267 or www.samswebsite.com

Prices subject to change.

GUIDE TO CABLING AND COMMUNICATION WIRING

Author: LOUIS COLUMBUS
ISBN: 0790612038 ● **SAMS#:** 61203
Pages: 320 ● **Category:** Communications
Case qty: TBD ● **Binding:** Paperback
Price: $39.95 US/$63.95CAN

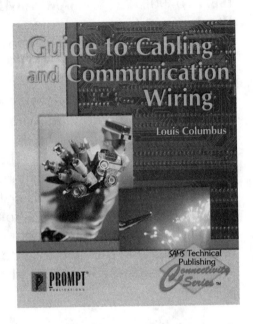

About the book: Part of Sams Connectivity Series, *Guide to Cabing and Communication Wiring* takes the reader through all the necessary information for wiring networks and offices for optimal performance. Columbus goes into LANs (Local Area Networks), WANs (Wide Area Networks), wiring standards and planning and design issues to make this an irreplaceble text.

PROMPT® Pointers:
Features planning and design discussion for network and telecommunications applications. Explores data transmission media. Covers Packet Framed-based data transmission.

Related Titles: *Administrator's Guide to E-Commerce*, by Louis Columbus, ISBN 0790611872. *Exploring LANs for the Small Business and Home Office*, by Louis Columbus, ISBN 0790612291. *Computer Networking for the Small Business and Home Office*, by John Ross, ISBN 0790612216.

Author Information: Louis Columbus has over 15 years of experience working for computer-related companies. He has published 10 books related to computers and has published numerous articles in magazines such as *Desktop Engineering, Selling NT Solutions*, and *Windows NT Solutions*. Louis resides in Orange, Calif.

To order today or locate your nearest PROMPT® Publications distributor at 1-800-428-7267 or www.samswebsite.com

Prices subject to change.

HOW THE PC HARDWARE WORKS

Author: MICHAEL GRAVES
ISBN: 079061250X ● **SAMS#:** 61250
Pages: 800 ● **Category:** Computer Technology
Case qty: TBD ● **Binding:** Paperback
Price: $39.95 US/$63.95CAN

About the book: As the technology surrounding our desktop PCs continues to evolve at a rapid pace, the opportunity to understand, repair and upgrade your PC is attractive. In an era where the PC you bought last year is now "out of date", your opportunity to bring your PC up-to-date rests in this informative text. Renouned author Michael Graves addresses this subject in a one-on-one manner, explaining each category of computer hardware in a complete, concise manner.

Prompt Pointers: Designed to bring a beginner up to a professional level of hardware expertise. Includes new SCSI III implementations, new video standards, and previews of upcoming technologies.

Related Titles: *Exploring Office XP,* ISBN 079061233X, *Designing Serial SANS,* ISBN 0790612461, *Administrator's Guide to Datawarehousing,* ISBN 0790612496.

Author Information: Michael Graves is a Senior Hardware Technician and Network Engineer for Panurgy of Vermont. Graves has taught computer hardware courses on the college level at Champlain College in Burlington, Vermont and The Essex Technical Center in Essex Junction, Vermont. While this is his first full-length book under his own name, his contributions have been included in other works and his technical writing has been the source of several of the more readable user's guides and manuals for different products.

To order today or locate your nearest PROMPT® Publications distributor at 1-800-428-7267 or www.samswebsite.com

Prices subject to change.

AUTOMOTIVE AUDIO SYSTEMS

Author: HOMER L. DAVIDSON
ISBN: 0790612356 • **SAMS#:** 61235
Pages: 320 • **Category:** Automotive
Case qty: TBD • **Binding:** Paperback
Price: $39.95 US/$63.95CAN

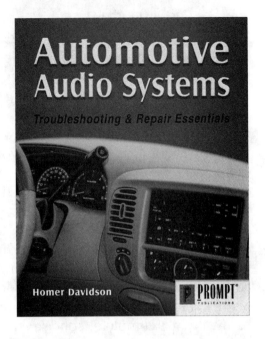

About the book: High-powered car audio systems are very popular with today's under-30 generation. These top-end systems are merely a component within the vehicle's audio system, much as your stereo receiver is a component of your home audio and theater system. Little has been written about the troubleshooting and repair of these very expensive automotive audio systems. Homer Davidson takes his decades of experience as an electronics repair technician and demonstrates the ins-and-outs of these very high-tech components.

Prompt Pointers: Coverage includes repair of CD, Cassette, Antique car radios and more. All of today's high-end components are covered. Designed for anyone with electronics repair experience.

Related Titles: *Automotive Electrical Systems*, ISBN 0790611422. *Digital Audio Dictionary*, ISBN 0790612011. *Modern Electronics Soldering Techniques*, ISBN 0790611996.

Author Information: Homer L. Davidson worked as an electrician and small appliance technician before entering World War II teaching Radar while in the service. After the war, he owned and operated his own radio and TV repair shop for 38 years. He is the author of more than 43 books for TAB/McGraw-Hill and Prompt Publications. His first magazine article was printed in *Radio Craft* in 1940. Since that time, Davidson has had more than 1000 articles printed in 48 different magazines. He currently is TV Servicing Consultant *for Electronic Servicing & Technology* and Contributing Editor for *Electronic Handbook*.

To order today or locate your nearest PROMPT® Publications distributor at 1-800-428-7267 or www.samswebsite.com

Prices subject to change.

DESIGNING SERIAL SANS

Author: WILLIAM DAVID SCHWADERER
ISBN: 0790612461 • **SAMS#:** 61246
Pages: 320 • **Category:** Computer Technology
Case qty: TBD • **Binding:** Paperback
Price: $39.95 US/$63.95CAN

About the book: The use of Serial SANS is an increasingly popular and efficient way to store data in a medium to large corporation setting. Serial SANS effectively stores your company's data away from the traditional server, allowing your valuable server resources to be used for running applications.

Prompt Pointers: Covers Device Specialization Considerations. Explains Media Signals, Data Encoding and Protocols. Discusses SAN hardware building blocks.

Related Titles: *Administrator's Guide to Datawarehousing*, ISBN 0790612496, *How the PC Hardware Works*, ISBN 079061250X.

Author Information: W. David Schwaderer has extensive complex computer system experience and was involved in the creationof two Silicon Valley start-up companies. Schwaderer has a diverse background in connectivity products, personal computer software, and voice DSP based systems. Schwaderer currently resides in Saratoga, Calif.

To order today or locate your nearest PROMPT® Publications distributor at 1-800-428-7267 or www.samswebsite.com

Prices subject to change.

ADMINISTRATOR'S GUIDE TO THE EXTRANET/INTRANET

Author: CONRAD PERSSON
ISBN: 0790612410 • **SAMS#:** 61241
Pages: 304 • **Category:** Computer Technology
Case qty: TBD • **Binding:** Paperback
Price: $34.95 US/$55.95CAN

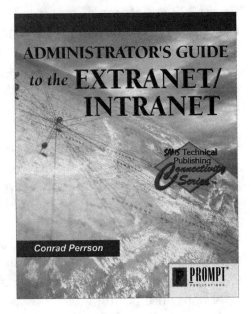

About the book: We are all familiar with the Internet, but few of us have occasion to utilize an Intranet or Extranet application. Both have vast applications related to inner-company communication, customer service, and vendor relations. Both are built similarl to Internet sites, and have many of the same features, issues, and problems. Intranet and Extranet applications are generally under-utilized, even though they provide the opportunity for both communication and financial benefits.

Prompt Pointers: Designed for the Systems Administrator or advanced webmaster. Outlines Intranet/Extranet issues, problems, and opportunities. Discusses hardware and software needs.

Related Titles: *Administrators Guide to E-Commerce*, ISBN 0790611872. *Computer Networking for the Small Business and Home Office*, ISBN 0790612216. *Exploring Microsoft Office XP*, ISBN 079061233X.

Author Information: Conrad Persson is the editor of *ES&T Magazine*, the premier publication for the electronics servicing industry. Conrad has decades of experience related to electronics and computer applications and resides in Shawnee Mission, Kan.

To order today or locate your nearest PROMPT® Publications distributor at 1-800-428-7267 or www.samswebsite.com

Prices subject to change.

BROADBAND EXPOSED

Author: MICHAEL BUSBY
ISBN: 0790612488 ● **SAMS#:** 61248
Pages: 352 ● **Category:** Communications
Case qty: TBD ● **Binding:** Paperback
Price: $39.95 US/$63.95CAN

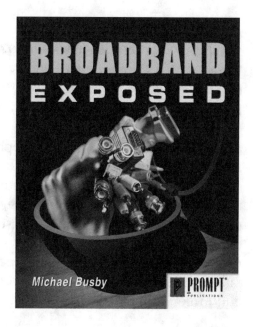

About the book: As the telecommunications industry goes through deregulation, the lines between telecom and other communication applications have become very blurred. Business mergers and advancing technology have created a need for more and more broadband applications. Author Michael Busby addresses these issues in relation to the telecommunications sector as well as topics pertaining to cable, satellite, RF, microwave and other communication methods. A must read for anyone working with telecommunication technologies.

Prompt Pointers: Includes discussions of LAN, CAD, imaging, wire, cable and more. Addresses networking fundamentals, protocols, and multimedia applications.

Related Titles: *Telecommunication Technologies*, ISBN 0790612259, *Exploring LANS for the Small Business & Home Office*, ISBN 0790612291, *Guide to Cabling & Communication Wiring*, ISBN 0790612038.

Author Information: Michael Busby is president and CEO of Mikal Enterprises, a global telecommunications design and consulting company. Busby has over 30 years telecommunications experience as field service engineer, systems engineer, R&D engineer, engineering manager, product manager, and VP engineering.

To order today or locate your nearest PROMPT® Publications distributor at 1-800-428-7267 or www.samswebsite.com

Prices subject to change.

GUIDE TO DEDICATED MICROPROCESSOR FUNDAMENTALS

Author: BOB ROSE
ISBN: 0790612402 • **SAMS#:** 61240
Pages: 320 • **Category:** Electronics Technology • **Binding:** Paperback
Price: $34.95 US/$55.95CAN

About the book: Dedicated Microprocessors are found in almost every consumer electronic device on the market today. Although not a repairable component, the dedicated microprocessor holds the clues to many successful repairs. This text discusses structure, function, communications, operating fundamentals, failures, requirements, peripheral problems and more.

Prompt Pointers: Dedicated microprocessors are found in most electronic devices. Technicians of all levels need to understand dedicated microprocessors to effectively repair devices. Little has been written on dedicated microprocessors related to repairs.

Related Titles: *Manufacturer to Manufacturer Part Number Cross Reference with CD-ROM*, ISBN 0790612321. *Semiconductor Cross Reference Guide 5E*, ISBN 0790611392. *DSP Filters*, ISBN 0790612046.

Author Information: Bob Rose has spent his career in the electronics servicing industry. An expert at the troubleshooting and repair of TVs and VCRs, Rose has and continues to service all the major brands of electronic devices. Bob Rose holds more than 40 training certificates and is the author of more than 30 articles and two books. As a member of the National Electronics Service Dealers Association, Bob stays abreast of the latest trends in technology from his home in Medina, Tenn.

To order today or locate your nearest PROMPT® Publications distributor at 1-800-428-7267 or www.samswebsite.com

Prices subject to change.

HOME THEATER SYSTEMS

Author: BOB GOODMAN
ISBN: 0790612372 • **SAMS#:** 61237
Pages: 304 • **Category:** Video Technology
Case qty: TBD • **Binding:** Paperback
Price: $39.95 US/$63.95CAN

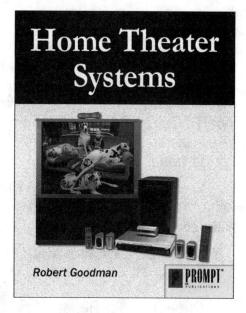

About the book: In days past, you had a TV, radio, and maybe a turntable in your "living room." Today, the evolution of electronics has brought us the Home Theater System, combining projection TVs, high-powered audio receivers, multiple CD players, DVD systems, surround-sound and more. This plethora of components is rarely purchased from a single manufacturer, making installation and maintenance a complicated task at best. Bob Goodman applies his electronics experience to this topic and provides a guidebook to home theater systems, including information on systems, components, troubleshooting, and maintenance.

Prompt Pointers: Home theater systems are the future of home audio/video systems. A buyer's guide is included. Great detail is provided regarding component choices.

Related Titles: *Digital Audio Dictionary*, ISBN 0790612011. *DVD Player Fundamentals*, ISBN 0790611945. *Guide to Satellite TV Technology*, ISBN 0790611767.

Author Information: Bob Goodman, CET, has devoted much of his career to developing and writing about more effective, efficient ways to troubleshoot electronics equipment. An author of more than 62 technical books and 150 technical articles, Goodman spends his time as a consultant and lecturer in Western Arkansas.

To order today or locate your nearest PROMPT® Publications distributor at 1-800-428-7267 or www.samswebsite.com

Prices subject to change.

Instructions for Using the Included CD-ROM

The CD-ROM included with this book includes device listings and data sheets for the Philips Semiconductors HEF4000 CMOS family. Data files for the CD-ROM have been provided courtesy of Philips Semiconductors. Before deciding to use a certain device, readers are asked to consult with the local Philips sales office or the nearest franchised distributor of Philips Semiconductors, because products may become obsolete due to no or low market demand.

Files included on the CD-ROM are in PDF format and are best viewed using Acrobat Reader 4.0 or higher. To load the latest version of Acrobat Reader free of charge, go to the following Web site:

http://www.samswebsite.com/photofact/efact.html

Once Acrobat Reader has been downloaded and is ready to use, insert the CD-ROM into your computer's CD-ROM drive. Start Acrobat Reader. When Acrobat Reader is opened, click on FILE and then OPEN. After clicking OPEN, find your CD-ROM drive directory and click on PHILIPS CMOS INDEX.PDF. This will open a list of device numbers. Clicking on the individual numbers will open data sheets for each device. In order to return to the main index page, click on the double, left-facing arrow at the top of the page until reaching the main page.

As a bonus for readers, a copy of the Sams Technical Publishing Fall Book Catalog also is included on the CD-ROM.